Energy: Management, Supply and Conservation

This book is dedicated to Prem, Lucy and Rachel

Energy: Management, Supply and Conservation

Dr Clive Beggs

BUTTERWORTH HEINEMANN

OXFORD AMSTERDAM BOSTON LONDON NEW YORK PARIS
SAN DIEGO SAN FRANCISCO SINGAPORE SYDNEY TOKYO

Butterworth-Heinemann
An imprint of Elsevier Science
Linacre House, Jordan Hill, Oxford OX2 8DP
225 Wildwood Avenue, Woburn MA 01801–2041

First published 2002

British Library Cataloguing in Publication Data
A catalogue record for this book is available from the British Library

Library of Congress Cataloguing in Publication Data
A catalogue record for this book is available from the Library of Congress

ISBN 0 7506 5096 6

For information on all Butterworth-Heinemann publications
visit our website at www.bh.com

Typeset by Integra Software Services Pvt. Ltd, Pondicherry, India
Printed and bound in Great Britain by Martins the Printers Ltd

Contents

1

Energy and the environment

Society in the developed world is built on the assumption that energy is both freely available and relatively cheap. However, there are environmental costs associated with the continued use of fossil fuels and these are causing a reappraisal of the way in which energy is used. This chapter investigates the global use of energy and its impact on the environment.

1.1 Introduction

Those of us who live in developed countries take energy very much for granted. Although we may not understand exactly what it is, we certainly know how to use it. Indeed, never before has there been a society which is so reliant on energy as our own. Consider for a moment the number of everyday items of equipment, tools and appliances that run on electricity; lamps, washing machines, televisions, radios, computers and many other 'essential' items of equipment all need a ready supply of electricity in order to function. Imagine what life would be like without electricity. Both our home and our working lives would be very different. Indeed, our high-tech computer reliant society would cease to function; our productivity would fall drastically; and our gross domestic product (GDP) would be greatly reduced; a fact highlighted by the power cuts that brought California to its knees in 2001 [1]. Similarly, if oil supplies ceased then the fabric of our society would very quickly fall apart. Those living in the UK may remember the events of September 2000, when a relatively small number of 'fuel protesters' managed to almost entirely stop petroleum supplies to the UK's petrol stations, with the result the economy came to a halt within days; people couldn't get to work and the supermarkets ran out of food. Those in the UK with longer memories might also recall how a combination of striking coal miners, power workers and crude oil price rises in the 1970s brought the UK to a standstill; electricity power cuts were commonplace, vehicle speed restrictions were introduced, and ultimately the government was forced to introduce a three-day working week in order to save energy. Clearly, although it is all too often taken for granted, cheap and available energy is essential to the running of any so-called advanced industrialized society. Understanding the nature of energy, its supply, and its utilization is therefore an important subject, for without it we in the developed world face an uncertain future.

To some of you reading this book, the society which has just been described may seem alien. Those living in developing countries will be all too aware that energy is a very finite resource. In many poorer countries, electricity is supplied only to major towns, and even then, power cuts are commonplace. This not only reduces the quality of life of those living in such countries, but also hampers productivity and ultimately ensures that those countries have a low GDP. If you live in one of these poorer nations you are in the majority, a majority of the world's population which consumes the minority of its energy. This is indeed, a great paradox. One-third of the world's population lives in a consumer society which squanders energy all too easily, while the other two thirds live in countries which are often unable to secure enough energy to grow economically. This is highlighted by the example of the USA which consumes approximately 26% of the all world's energy [2], while having only 4.4% of the world's population.

The inequalities between developed and developing countries are real and should be cause for great concern to the whole world. Unfortunately, political self-interest is often much stronger than altruism and the gap between the rich and the poor nations has widened in recent years. However, when confronted with unpalatable facts about gross inequalities between rich and poor nations, our usual response is to assume that the problem is altogether too large to solve, and to forget about it. After all most of us have many other pressing needs and problems to worry about. This of course, is a very understandable response. However, forgetting about the problem does not mean that it will go away. In fact, the reality is that as the economies of the developing world grow, so will their demand for energy [3]. This will increase pressure on the Earth's dwindling supply of fossil fuel and will also increase greenhouse gas emissions and atmospheric pollution in general. It is worth remembering that the Earth is a relatively small place and that atmospheric pollution is no respecter of national boundaries. Indeed, issues such as climate change and third world debt are now impinging on the comfort and security of the developed world. It is the perceived threat of global climate change that has been the driving force behind all the intergovernmental environmental summits of the late twentieth century. In historical terms the summits at Montreal, Rio and Kyoto were unique, since never before had so many nations sat down together to discuss the impact of humans on the environment. Indeed, it could truthfully be said that never before in the history of the world have so many sat down together to discuss the weather! Collectively these summits produced protocols which set targets for reducing ozone depleting and greenhouse gas emissions, and have forced governments around the world to reappraise policies on energy supply and consumption. The collective agreements signed at these summits have impacted, to varying degrees, on the signatory nations and have manifest themselves in a variety of ways. For example, in the UK a large proportion of the electricity supply sector has switched from coal, which has a high carbon intensity, to natural gas, which has a much lower carbon content. In the construction industry, so-called 'green buildings' are being erected which are passively ventilated and cooled (see Chapter 13) with the express intention of minimizing energy consumption and eliminating the use of harmful refrigerants. In addition, the high profile nature of the various intergovernmental summits has meant that concern about energy and its utilization is now at the forefront of public consciousness.

Because most lay people focus on the consumption of energy it is often forgotten that the supply of energy is itself a large and important sector of the world's economy. For example, the energy industry in the UK is worth 5% of GDP and employs 4% of the industrial workforce (1999 data) [4], making it one of the largest industries in the UK. The energy supply sector is also very multinational in nature. For example, crude oil is transported all around the globe, with a total of 41 048 barrels being transported daily in 1999 alone [2]. Similarly, large quantities of natural gas are piped daily over long distances and across many international borders, and electricity is traded between nations on a daily basis. Given the size of the energy supply industry, its multinational nature, and its importance to the world economy, it should come as no surprise that many parties have a vested interest in promoting energy consumption and that this often leads to conflict with those driven by environmental considerations.

1.2 Politics and self-interest

Any serious investigation of the subject of energy supply and conservation soon reveals that it is impossible to separate the 'technical' aspects of the subject from the 'politics' which surround it. This is because the two are intertwined; an available energy supply is the cornerstone of any economy and politicians are extremely interested in how economies perform. Politicians like short-term solutions and are reluctant to introduce measures which will make them unpopular. Also, many political parties rely on funding from commercial organizations. Consequently, political self-interest often runs counter to collective reason. For example, in many countries (although not all), politicians who put forward policies which promote congestion charging or petrol price increases become unpopular, and are soon voted out of office. As a result, measures which might at first sight appear to be extremely sensible are discarded or watered down due to political self-interest. It is of course far too easy to blame politicians for hypocrisy, while ignoring the fact that we as individuals are also often culpable. Consider the case of a rapidly growing large city which has traffic congestion problems; journey times are long and air quality is poor. Clearly the quality of life of all those in the city is suffering due to the road congestion. The solution is obvious. People need to stop using their cars and switch to public transport. If questioned on the subject, car drivers will probably agree that the city is too congested and that something should be done to reduce the number of cars on the roads. However, when it is suggested that they, as individuals, should stop using their own cars then self-interest tends to win over reason; objections are raised, sometimes violently, that such a measure is too extreme and that the freedom of the individual is being compromised. From this we can only conclude that it is impossible for politicians alone to bring about change in 'energy politics' without changes in public opinion. In many ways it is true to say we all get the leaders we deserve!

The road congestion example discussed above is a good illustration of the contradiction between reason and self-interest, which is often manifest within the individual. However, exactly the same contradiction is often all too evident at a governmental and international level. When it comes to environmental issues governments often refuse to implement sound policies because in so doing they might inhibit economic growth. To those concerned with environmental issues, the idea of putting national 'self-interest' before the environmental health of the planet might seem absurd. However, the issue is not as clear-cut as it would appear at first sight. There is a strong link between energy consumption and GDP. Without a cheap and available energy supply the economic growth of many nations will be restricted. Consequently, any enforced reduction in GDP due to environmental control measures is going to be much more painful to the inhabitants of poorer countries than an equivalent cut in a developed country. Indeed, to many poorer nations, the notion of rich developed countries telling them to reduce greenhouse gas emissions is hypocritical; after all, the advanced nations of North America and western Europe only became rich through intensive manufacturing. Since the eighteenth century the developed countries have consumed large amounts of primary energy and produced high levels of pollution. So in the late twentieth century when – having created many environmental problems – these same nations turn to their poorer neighbours and expect them to restrict economic growth in the name of environmentalism, it is not surprising that to many in the developing world this approach appears high-handed. Therefore, it is up to those of us in the developed world to lead by example and alter our approach towards energy consumption.

1.3 What is energy?

Before discussing global energy production and consumption, it is perhaps wise to go over some of the basic physics associated with the study of energy. We are all familiar with the term *energy*, but surprisingly few people fully appreciate its true nature. In everyday language the word *energy* is used very loosely; words like *work*, *power*, *fuel* and *energy* are often used interchangeably and frequently incorrectly. To the physicist or engineer, *energy* is a very specific term which is perhaps the best explained by means of an illustration.

Consider a mass of 1 kg which is raised 1 m above a surface on which it was originally resting. It is easy to appreciate that in order to raise the weight through the distance of 1 m, someone, or some machine, must have performed some work. In other words work has been put into the system to raise the mass from a low level to a higher level. This work is the amount of energy that has been put into the system. So, when the weight is in the raised position, it is at a higher energy level than when on the surface. Indeed, this illustration forms the basis for the International System (SI) unit of energy, the 'joule', which can be defined as follows.

> *One joule (J) is the work done when a force of 1 newton (N) acts on an object so that it moves 1 metre (m) in the direction of the force.*

and

> *One newton (N) is the force required to increase or decrease the velocity of a 1 kg object by 1 m per second every second.*

The number of newtons needed to accelerate an object can be calculated by:

$$F = m \times a \qquad (1.1)$$

where m is the mass of the object (kg), and a is the acceleration (m/s^2). Given that the acceleration due to gravity is 9.81 m/s^2, a mass of 1 kg will exert a force of 9.81 N (i.e. 1 kg $\times$ 9.81 m/s^2). Therefore the energy required to raise it through 1 m will be 9.81 J.

If the 1 kg mass is released it will fall through a distance of 1 m back to its original position. In doing so the *potential energy* stored in the 1 kg mass when it is at the higher level will be released. Notice that the energy released will be equal to the work put into raising the weight. For this reason the term *work* is sometimes used instead of *energy*. Perhaps a good way of viewing energy is to consider it as stored work. Therefore, *potential energy* represents work that has already been done and stored for future use. *Potential energy* can be calculated by:

$$\text{Potential energy} = m \times g \times h \qquad (1.2)$$

where m is the mass of the object (kg), g is the acceleration due to gravity (i.e. 9.81 m/s^2), and h is the height through which the object has been raised (m).

As the weight falls it will possess energy because of its motion and this is termed *kinetic energy*. The kinetic energy of a body is proportional to its mass and to the square of its speed. *Kinetic energy* can be calculated by:

$$\text{Kinetic energy} = 0.5 \times m \times v^2 \qquad (1.3)$$

where v is the velocity of the object (m/s)

We can see that during the time the mass takes to fall, its *potential energy* decreases whilst its *kinetic energy* increases. However, the sum of both forms of energy must remain constant during the fall. Physicists and engineers express this constancy in the 'law of conservation of energy', which states that the total amount of energy in the system must always be the same.

It should be noted that the amount of energy expended in raising the weight is completely independent of the time taken to raise the weight. Whether the weight is raised in 1 second or 1 day makes no difference to the energy put into the system. It does however, have an effect on the 'power' of the person or machine performing the work. Clearly, the shorter the duration of the lift, the more powerful the lifter has to be. Consequently, *power* is defined as the rate at which work is done, or alternatively, the rate of producing or using energy. The SI unit of power is the watt (W). Therefore, a machine requires a power of 1 watt if it uses 1 joule of energy in 1 second (i.e. 1 watt is 1 joule per second). In electrical terms 1 watt is the energy released in 1 second by a current of 1 ampere passing through a resistance of 1 ohm.

It is well known that if two rough surfaces are rubbed together, the work required in overcoming the friction produces heat. Also, it is known that electricity can be used to perform mechanical work by utilizing an electric motor. Therefore, it is clear energy can take a number of forms (e.g. electrical energy, mechanical work and heat) and it can be easily converted between these various forms. For example, a fossil fuel can be burnt to produce heat energy in a power station. The heat energy produced is then converted to mechanical energy by a turbine, which in turn produces electrical energy through a generator. Finally, the electricity is distributed to homes and factories where it can be converted to mechanical work using electric motors, heat using resistance elements and light using electric lamps.

1.3.1 Units of energy

For a myriad of reasons (too numerous to mention here), a bizarre array of units for energy has evolved. Books, articles and papers on energy quote terms such as 'kWh', 'therms', 'joules', 'calories', 'toe' and many more. This makes things very complicated and confusing for the reader. This section is, therefore, included to introduce some of the units more commonly in use.

Kilowatt-hour (kWh)

The kilowatt-hour (kWh) is a particularly useful unit of energy which is commonly used in the electricity supply industry and, to a lesser extent, in the gas supply industry. It refers to the amount of energy consumed in 1 hour by the operation of an appliance having a power rating of 1 kW. Therefore:

$$1 \text{ kWh} = 3.6 \times 10^6 \text{ joule}$$

British thermal unit (Btu)

The British thermal unit (Btu) is the old imperial unit of energy. It is still very much in use and is particularly popular in the USA.

$$1 \text{ Btu} = 1.055 \times 10^3 \text{ joule}$$

Therme

The therme is a unit that originated in the gas supply industry. It is equivalent to 100 000 Btu.

$$1 \text{ therme} = 1.055 \times 10^8 \text{ joule}$$

Tonne of oil equivalent (toe)

The 'tonne of oil equivalent' (toe) is a unit of energy used in the oil industry.

$$1 \text{ toe} = 4.5 \times 10^{10} \text{ joule}$$

Barrel

The barrel is another unit of energy used in the oil industry. There are 7.5 barrels in 1 toe.

$$1 \text{ barrel} = 6 \times 10^9 \text{ joule}$$

Calorie

In the food industry the calorie is the most commonly used unit of energy. It is in fact the amount of heat energy required to raise 1 gram of water through 1 °C.

$$1 \text{ calorie} = 4.2 \times 10^3 \text{ joule}$$

1.3.2   The laws of thermodynamics

Thermodynamics is the study of heat and work, and the conversion of energy from one form into another. There are actually three laws of thermodynamics, although the majority of thermodynamics is based on the first two laws.

The first law of thermodynamics

The first law of thermodynamics is also known as the law of conservation of energy. It states that the energy in a system can neither be created nor destroyed. Instead, energy is either converted from one form to another, or transferred from one system to another. The term 'system' can refer to anything from a simple object to a complex machine. If the first law is applied to a heat engine, such as a gas turbine, where heat energy is converted into mechanical energy, then it tells us that no matter what the various stages in the process, the total amount of energy in the system must always remain constant.

The second law of thermodynamics

While the first law of thermodynamics refers to the quantity of energy that is in a system, it says nothing about the direction in which it flows. It is the second law which deals with the natural direction of energy processes. For example, according to the second law of thermodynamics, heat will always flow only from a hot object to a colder object. In another context it explains why many natural processes occur in the way they do. For example, iron always turns to rust; rust never becomes pure iron. This is because all processes proceed in a direction which increases the amount of disorder, or chaos, in the universe. Iron is produced by smelting ore in a foundry, a process which involves the input of a large amount of heat energy. So, when iron rusts it is reverting back to a 'low energy' state. Although it is a difficult concept to grasp, disorder has been quantified and given the name 'entropy'. Entropy can be used to quantify the amount of useful work that can be performed in a system. In simple terms, the more chaotic a system, the more difficult it is to perform useful work.

In an engineering context it is the second law of thermodynamics that accounts for the fact that a heat engine can never be 100% efficient. Some of the heat energy from its fuel will be transferred to colder objects in the surroundings, with the result that it will not be converted into mechanical energy.

The third law of thermodynamics

The third law of thermodynamics is concerned with absolute zero (i.e. $-273\ °C$). It simply states that it is impossible to reduce the temperature of any system to absolute zero.

1.4 Energy consumption and GDP

In the introduction to this chapter it was stated that it is impossible to remove politics from any discussion or study of energy. This is because the GDP of any nation is related to its energy consumption. Perhaps the best way to illustrate this link is to look at energy consumption from an historical viewpoint. Table 1.1 shows the estimated average daily consumption of people in various historical societies.

It can be seen from Table 1.1 that per capita energy consumption has increased (almost exponentially) as societies have become more advanced and industrialized. The first humans were simple gatherers who lived off wild fruit, nuts and vegetables. However, as people began to hunt and live in less hospitable regions, they learnt to use fire for cooking and heating. As time progressed, societies developed; first came agriculture and then came industrial practices; the smelting and working of metals and increased trading of goods and materials. With these technological and social advances came increased energy consumption; buildings needed heating, food needed cooking and manufacturing processes required fuel. It is estimated that per capita energy consumption rose from approximately 4000 kilocalories per day, in the age of the hunter-gatherer, to approximately 21 000 kilocalories per day, in Europe prior to the Industrial Revolution [5]. The Industrial Revolution, first in Europe and later in North America, resulted in a rapid increase in per capita energy consumption during the nineteenth century. Populations

Table 1.1   Historical overview of per capita energy consumption [5]

Period and location	Type of society	Characteristics	Estimated daily per capita energy consumption
Very early	Gatherers	Gathered wild fruit, nuts and vegetables	2 000 kCal (8.2 MJ)
1000 000 BC	Hunter-gatherers	Gathered wild fruit etc., hunted and cooked food	4 000 kCal (16.4 MJ)
4000 BC Middle East	Settled farmers	Sowed crops and kept animals	12 000 kCal (49.2 MJ)
AD 1500 Europe	Agricultural with small scale industry	Agricultural society with specialized industries producing metal, glass etc.	21 000 kCal (88.2 MJ)
AD 1900 Europe	Industrialized society	Large scale industry, mass production, large cities	90 000 kCal (378 MJ)
AD 1990 USA, Western Europe etc.	Advanced industrialized society	Consumer society, mass transport, many labour saving devices	250 000 kCal (1 GJ)

grew rapidly and became concentrated in large towns and cities. Mass production became commonplace and with it more transportation of goods, raw materials and people. This dramatic increase in energy consumption continued throughout the twentieth century as more and more societies became industrialized, to such an extent that in technologically advanced countries such as the USA, per capita energy consumption has reached approximately 250 000 kilocalories per day [5].

From the above historical review, it is clear that there is a strong link between per capita energy consumption and economic growth. In simple terms, less developed agrarian societies consume much less energy than their advanced industrial counterparts. Figures 1.1 and 1.2 show data derived from the World Development Report for 1990 produced by the World Bank [6]. These figures illustrate the relationship between per capita gross national product (GNP) (which is closely related to GDP) and energy consumption for some of the world's poorer and richer nations, respectively. Although energy consumption is influenced by factors such as population density, weather and location, it can be seen from Figures 1.1. and 1.2 that for most nations, particularly developing countries, there is still a strong correlation between GNP and energy consumption; broadly speaking in most societies energy consumption and economic growth tend to move in parallel.

Although there has been a strong historical link between GDP and energy consumption, in recent years there has been a decoupling of the relationship in many of the more advanced countries. It has been observed that since the 1970s in these countries, increased GDP has not been accompanied by a pro-rata increase in energy consumption. Indeed, in the UK and a number of other European countries, energy consumption has plateaued and remained relatively constant over recent years [5]. This phenomenon is occurring partly because of the adoption of newer energy efficient technologies and partly because many older energy intensive manufacturing industries are being replaced by high-tech and service sector industries, which consume much less energy. However, from a global perspective, it is simplistic to argue that this move towards the service sector is conserving energy, since in reality these countries are effectively

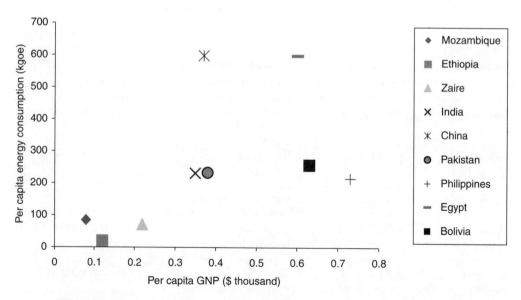

Figure 1.1   Per capita GNP versus energy consumption of some of the poorer nations[6]

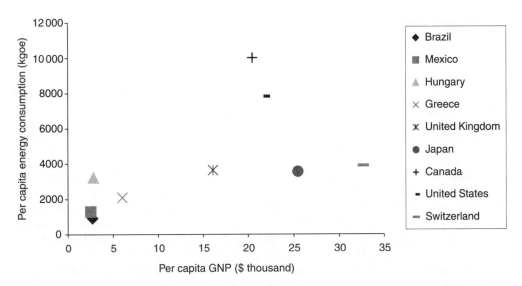

Figure 1.2 Per capita GNP versus energy consumption of some of the richer nations[6]

exporting their manufacturing and heavy industry requirements to other parts of the world where wage costs are lower. Indeed, there is evidence that many advanced 'consumer' nations are simply exporting their 'dirty' energy intensive industries to countries in which environmental legislation is much weaker, with the result that in gross terms environmental pollution is increasing.

The ratio of energy used to GDP is known as the *energy intensity* of an economy. It is a measure of the output of an economy compared with its energy inputs, in effect a measure of the efficiency with which energy is used. Manufacturing nations, with old or relatively poor infrastructures, like many of the East European and former Soviet Union (FSU) countries, often exhibit very high *energy intensities*, while the more energy efficient 'post-industrialized' nations have much lower intensities. The link between infrastructure and *energy intensity* is very strong indeed [7]. In developing countries, development of an infrastructure leads to growth in energy intensive manufacturing industries. In industrialized economies, *energy intensity* is strongly influenced by the efficiency of the infrastructure and capital stock such as power stations, motor vehicles, manufacturing facilities and end-user appliances. The energy efficiency of capital stock is, in turn, influenced by the price of energy relative to the cost of labour and the cost of borrowing capital. If energy costs are high in relation to these other costs, then it is much more likely that investments will be made in energy-efficient technologies. Conversely, if energy prices are low, then little incentive exists for investment, or indeed research, in more energy efficient technologies [7].

While energy intensity is strongly influenced by the price of energy, it is also affected by factors which are not directly attributable to price effects. For example, changes in technology and changes in the composition of world trade can influence energy intensity. Geographical location has a strong influence; cold northerly countries tend to exhibit high energy intensities. Other factors include changes in fashion and preferences. For example, if the practice of cycling to work becomes popular with enough people, then it is possible that this will influence the energy intensity of an economy. In short, there are many factors which influence energy intensity.

1.5 Environmental issues

A full examination of the environmental problems facing the Earth, although very interesting, is well beyond the scope of this book. However, because environmental considerations, in particular the perceived threat of global warming, are influential in shaping the energy policy of many countries, it is essential that the issue be discussed in some detail. It is the perceived threat of climate change, above any other issue, which is changing the attitudes towards energy consumption. Although there is much scientific debate on the precise nature and extent of the twin threats of global warming and ozone depletion, the fact remains that these threats, whether actual or imaginary, are perceived by many governments to be real, with the result that both national and international energy policies are now being driven by an environmental agenda. It is important to have an understanding of pertinent environmental issues. Ignorance of the facts relating to environmental issues is surprisingly widespread amongst politicians, professionals and the public at large. Concepts such as global warming and ozone depletion are often confused and interchanged. Indeed, some individuals committed to environmentally *green* lifestyles exhibit very woolly thinking when it comes to the science of the environment. This section is, therefore, written with the sole intent of presenting the relevant facts and explaining the pertinent issues relating to global warming and ozone depletion.

1.5.1 Global warming

There is growing scientific evidence that greenhouse gas emissions caused by human activity are having an effect on the Earth's climate. The evidence suggests that the Earth's climate has warmed by almost 0.7 °C since the end of the nineteenth century [8], and that the pace of this warming is increasing. Globally, the 1990s were the warmest years on record, with seven of the ten warmest years being recorded in that decade [8, 9]. Indeed, in 1998 the global temperature was the highest since 1860 [8] and was the twentieth consecutive year with an above normal global surface temperature [8]. Figure 1.3 illustrates the steady rise in global temperature.

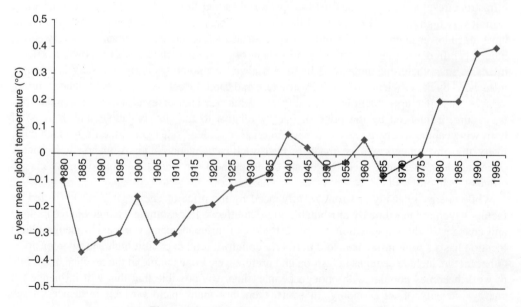

Figure 1.3   Global mean temperature change[9]

It is generally accepted that the rapid rise in global temperature experienced during the latter part of the twentieth century is due, in part, to atmospheric pollution arising from human activity, which is accelerating the Earth's greenhouse effect. The greenhouse effect is a natural phenomenon which is essential for preserving the 'warmth' of the planet. It is caused by trace gases in the upper atmosphere trapping long-wave infrared radiation emitted from the Earth's surface. The Earth's atmosphere allows short-wave solar radiation to pass relatively unimpeded. However, the long-wave radiation produced by the warm surface of the Earth is partially absorbed and then re-emitted downwards by greenhouse gases in the atmosphere. In this way an energy balance is set up, which ensures that the Earth is warmer than it would otherwise be. Without the greenhouse effect it is estimated that the Earth's surface would be approximately 33 °C cooler [10], and almost uninhabitable. Although the greenhouse effect is essential to the well being of human populations, if greenhouse gas levels rise above their natural norm, the consequent additional warming could threaten the sustainability of the planet as a whole.

The main naturally occurring greenhouse gases in the Earth's atmosphere are water vapour and CO_2. Of these, it is water vapour which has the greatest greenhouse action. While CO_2 concentrations are strongly influenced by human activity, atmospheric water vapour is almost entirely determined by climatic conditions and not human action. Human activity is responsible for production of a number of other potent greenhouse gases, including methane, nitrous oxide, chlorofluorocarbons (CFCs) and hydrochlorofluorocarbons (HCFCs). From the late eighteenth century onwards, concentrations of 'man-made' greenhouse gases (with the exception of CFCs and HCFCs, which were first introduced in the 1930s) have steadily increased. Table 1.2 shows the pre-industrial and 1990 levels of various greenhouse gases. For each gas it can be seen that there has been a substantial rise in the atmospheric concentration. For example, CO_2 concentrations have grown from 280 ppm in the middle of the eighteenth century, to approximately 353 ppm in 1990: a rise of about 26%, leading to a current rate of increase of about 0.5% a year [10]. Indeed, the Intergovernmental Panel on Climate Change (IPCC) forecast that a likely doubling of atmospheric CO_2 will occur by 2050, leading to an average global temperature increase of between 1.5 °C and 4.5 °C [10, 11].

Although CO_2 is the single 'man-made' gas which contributes most towards overall global warming (i.e. in excess of 50%), it is by no means the most potent of the greenhouse gases. Methane for example, is approximately 21 times as potent as CO_2. In other words, methane has a relative global warming potential (GWP) of 21 compared with that of CO_2, which is 1. Incredibly, CFC-11 has a GWP of approximately 3500 and CFC-12 has a GWP of approximately 7300 [10], making CFCs the most potent of greenhouse gases. CFCs were first introduced in the 1930s and were widely used as refrigerants, solvents and aerosol propellants, until they were

Table 1.2   Contribution to global warming of various gases [12]

Greenhouse gas	Carbon dioxide equivalent per molecule	Pre-1800 concentration	1990 Concentration	Growth rate (%/year)	Atmospheric life (years)
Carbon dioxide	1	280 ppmv	353 ppmv	0.50	50–200
Methane	21	0.8 ppmv	1.72 ppmv	0.90	10
CFC 12	7300	0.0 ppmv	484 pptv	4.00	130
CFC 11	3500	0.0 ppmv	280 pptv	4.00	65
Nitrous oxide	290	288 ppbv	310 ppbv	0.25	150

withdrawn in the mid-1990s. They are very stable and remain in the upper atmosphere for considerable periods of time, as much as 130 years in the case of CFC-12 [10]. Given that they are also potent ozone depletors, it is not surprising that the control and elimination of CFCs became one of the major environmental targets in the 1990s.

The extent to which global warming is likely to occur as a result of the build-up of greenhouse gases is a matter of much scientific debate. The Hadley Centre of the UK Meteorological Office predicts that, under the 'business as usual' scenario, the world's climate will warm by about 3 °C over the next 100 years [12], which is in keeping with the IPCC's forecast of a 1.5 °C to 4.5 °C rise by 2050 [10]. Although there is general agreement that climate change is the most serious environmental threat facing the world today, the precise nature of this 'climate change' is open to debate. It is predicted that as global warming progresses, sea levels will rise by over 400 mm by 2080 [12] due to the combined effects of thermal expansion of the oceans and melting of polar ice. This will put the lives of millions of people at risk, with an additional 80 million people particularly threatened with flooding in the low lying parts of southern and South-East Asia [12]. It is probable that droughts will occur due to increased temperatures and that Africa, the Middle East and India will all experience significant reductions in cereal crop yields [12]. Increased drought will mean that by 2080 an additional 3 billion people could suffer increased water stress, with Northern Africa, the Middle East and the Indian subcontinent expected to be the worst affected [12]. It is ironic that it will be the poorest countries, often ones which have contributed the least to global warming, which are most likely to be vulnerable to the effects of climate change.

1.5.2 Carbon intensity of energy supply

Carbon intensity is a measure of the amount of CO_2 that is released into the atmosphere for every unit of energy produced. As such it is wholly dependent on the type of fuel used. For example, electricity produced from nuclear power plants produces no CO_2 emissions, whereas that produced from coal-fired power station has a high carbon intensity. Table 1.3 shows the relative carbon intensities for electricity produced from a variety of fuels.

While renewable energy sources such as wind, solar, and hydropower emit no CO_2, the carbon content of fossil fuels varies greatly. It can be seen from Table 1.3 that electricity produced in a typical coal-fired power station produces approximately 2.4 times as much CO_2 as that produced by a combined cycle gas turbine (CCGT) plant [13]. Indeed, it has been demonstrated that the carbon intensity of delivered mains electricity is not constant, but varies considerably with time and with the generation plant mix [14]. For example, in England and Wales, carbon intensity is at its lowest during the nighttime in summer, when the bulk of the power is produced from nuclear energy.

Table 1.3 CO_2 emissions per kWh of delivered electrical energy (compiled from Building Research Establishment data) [11]

Primary fuel	Kg of CO_2 per GJ of primary energy (kg/GJ)	Average gross efficiency of power plant (%)	Kg of CO_2 per GJ of delivered electrical energy (kg/GJ)	Kg of CO_2 per kWh of delivered electrical energy (kg/kWh)
Coal	90.7	35	259.1	0.93
Oil	69.3	32	216.6	0.78
Gas (CCGT)	49.5	46	107.6	0.39

It is possible to achieve significant reductions in carbon intensity simply by switching from a fuel such as coal, which has a high carbon intensity, to one with a much lower intensity, such as natural gas. This is in fact what happened in the UK during the 1990s when there was a massive switch from coal to natural gas as the fuel of choice for electricity generation.

Because carbon intensity is wholly dependent on the type of fuel used, it differs across regions and also over time. During the 1990s coal became less important as a source of energy in western Europe, with the shutting down of lignite production in Germany and of hard coal production in the UK [7]. For example, in England and Wales the switch from coal to natural gas which accompanied deregulation of the electricity supply industry meant that coal consumption dropped from 65 million tonnes of oil equivalent (mtoe) in 1989 to only 35.6 mtoe in 1999 [2]. This has resulted in a 45% decrease in the UK's carbon intensity from 1980 to 1998 [15]. By contrast in the USA during the 1990s the electricity generators continued to use coal extensively and as a result the carbon intensity for western Europe has dropped below that of North America in recent years [7].

1.5.3 Carbon dioxide emissions

It has been estimated that global CO_2 emissions will reach approximately 9.8 billion tonnes per annum by 2020 – 70% above 1990 levels [3]. In the industrialized world it is predicted that CO_2 emissions will increase from a 1990 level of 2.9 billion tonnes to 3.9 billion tonnes in 2020 (see Figure 1.4) [3]. Emissions are, however, predicted to grow at a slower rate than primary energy consumption, mainly because of growth in the use of natural gas relative to coal in the developed world. However, in the industrialized countries oil is predicted to remain the dominant source of CO_2 emissions because of increased automobile use.

In 1990 emissions from the developed countries were approximately twice as much as those of the developing world [3]. By 2010 it is predicted that CO_2 emissions in the developing countries will surpass those of the industrialized countries (see Figure 1.4) [3]. The rapid growth in CO_2 emissions from the developing world is predicted because high rates of economic growth are anticipated and also because of the heavy dependence on coal in the developing Asian

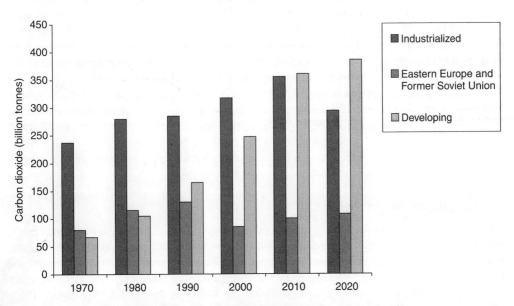

Figure 1.4 World carbon dioxide emissions by region[3]

countries. Predictions indicate that China and India alone will account for the majority of the worldwide increase in coal consumption by 2020 [3].

1.5.4 Depletion of the ozone layer

Ozone (O_3) in the Earth's stratosphere performs the vital function of protecting the surface of the planet from ultraviolet (UV) radiation which would otherwise be extremely harmful to human and animal life. The stratosphere is a layer approximately 35 km thick which has its lower limit at an altitude of 8–16 km. Ozone is produced in the stratosphere by the absorption of solar UV radiation by oxygen molecules (O_2) to produce ozone through a series of complex photochemical reactions [16]. The ozone produced absorbs both incoming solar UV radiation and also outgoing terrestrial long-wave radiation. In doing so, the ozone in the stratosphere is converted back to oxygen. The process is, therefore, both continuous and transient, with ozone continually being created and destroyed. The process is dependent on the amount of solar radiation incident on the Earth; consequently, ozone levels in the stratosphere are strongly influenced by factors such as altitude, latitude and season.

In the late 1970s a 'hole' was first discovered in the ozone layer above Antarctica [17]. Observations over a number of decades reveal that each September and October up to 60% of the total ozone above Antarctica is depleted [17]. In addition, progressive thinning of stratospheric ozone in both the northern and the southern hemispheres has been observed, with record low global ozone levels being recorded in 1992 and 1993 [17]. The average ozone loss across the globe has totalled about 5% since the mid-1960s, with the greatest losses occurring each year in the winter and spring [17]. This degradation of the ozone layer has resulted in higher levels of UV radiation reaching the Earth's surface. Increased UV radiation in turn leads to a greater incidence of skin cancer, cataracts, and impaired immune systems.

Blame for the recent and rapid deterioration of the ozone layer has been placed on escaping gases such as CFCs and nitrous oxide. Until recently CFCs were widely used in many applications including aerosol propellants, refrigerants, solvents and insulation foam. CFCs, especially CFC-11 and CFC-12, as well as being strong greenhouse gases are also potent ozone depletors. The lifetime of CFC-11 in the stratosphere is about 65 years, while that for CFC-12 is estimated to be 130 years [10]. In recent years, intergovernmental agreements, particularly the Montreal Protocol (1987), have phased out the production and use of CFCs. However CFCs are very long lived in the stratosphere and hence any reduction in CFC release will have little effect in the near future. The phasing out of CFCs has caused a greater reliance on HCFCs, in particular HCFC-22, which although much more ozone friendly is still a potent greenhouse gas. Under the Montreal Protocol HCFC-22 is being phased out and as a result the chemical companies are developing new generations of ozone 'friendly' refrigerants. Ozone depletion has caused many designers of buildings in Europe to question the need for vapour compression refrigeration machines to air condition buildings, with the result that alternative passive ventilation strategies are now being adopted in many new buildings (see Chapter 13).

1.5.5 Intergovernmental action

In the 1980s governments around the world became aware of some of the environmental problems associated with atmospheric pollution and the first in a series of intergovernmental summits was held in a concerted effort to combat the perceived problems. In many ways the Montreal Protocol, signed in September 1987, marks a turning point in global environmental policy. The leading industrialized nations signed the Montreal Protocol in September 1987, with the aim of limiting emissions of certain ozone depleting gases, such as CFCs (i.e. CFC-11,

CFC-12, CFC-113, CFC-114 and CFC-115) and halons (i.e. Halon 1211, Halon 1301 and Halon 2402). The original intention of the Protocol was to reduce consumption of these ozone degrading gases by 50% below the 1986 level by 1999 [16]. Since its original signing the Protocol has been reviewed regularly and such has been the concern about ozone depletion that the Protocol was expanded to cover HCFCs (i.e. HCFC-141b, HCFC-142b and HCFC-22) and the phase out schedule was also accelerated. In 1992, the parties to the Protocol agreed to accelerate the 100% phase out of CFCs, carbon tetrachloride, and methyl chloroform to the end of 1995 and halons to the end of 1993 [18]. The parties to the Protocol also agreed to phase out HCFCs so that a 90% reduction in production would be achieved by 2010 and a complete phase out by 2030 [19].

Having made a concerted effort to tackle the problem of ozone depletion the world's leading industrialized nations then turned their attention to the problem of global warming. Through a series of summits, notably Rio in 1992 and Kyoto in 1997, the nations formulated an international framework for reducing global CO_2 and other greenhouse gas emissions. At the Kyoto conference in December 1997, attended by 160 countries, the so-called 'Annex I' countries (which included the USA, Canada, the European Union (EU) countries, Japan, Australia, and New Zealand) agreed to reduce their emissions of six greenhouse gases (i.e. CO_2, nitrous oxide, methane, hydrofluorocarbon gases (HFCs), perfluorocarbons (PFCs) and sulphur hexafluoride) by at least 5% compared with 1990 levels between 2008 and 2012 [20]. However, the EU aims to reduce its emissions of the main six greenhouse gases (from 1990 levels) by 8% by 2012 [21, 22]. In order to meet this target the EU member states have taken various steps, including tightening building regulations and the introduction of carbon taxes.

1.5.6 Carbon credits and taxes

In order to meet their obligations under the Kyoto agreement a number of countries, notably The Netherlands, Sweden, Finland, Norway, Denmark and the UK have introduced 'carbon taxes'. These taxes are designed to penalize high carbon intensity energy consumption and promote the use of renewable energy sources. For example, the UK introduced its Climate Change Levy in 2001 with the express intention of increasing the share of its electricity generated by renewables from 2% to 10% by 2010 [15]. Nevertheless, many economists are sceptical about the use of carbon taxes, preferring instead a system of tradable emission permits. Under the Kyoto Protocol, flexibility was introduced into the agreement through 'Kyoto mechanisms' which allow countries to partake in emissions trading. It is argued that tradable permits are superior to carbon taxes, because unlike carbon taxes, they are a form of rationing which should ensure that targets are achieved. Permits are also more applicable to the international nature of the problem, since a regime of international carbon taxes would be extremely difficult to enforce.

The concept of trading in greenhouse gas emissions may seem very strange to many, so perhaps an analogy would be helpful at this point. Consider the case of a home-owner who wakes up one morning to find that a pipe has burst and that a flood has occurred [23]. Imagine also that this particular home-owner runs a profitable law firm and is also very good at mending burst pipes. The homeowner is, therefore, faced with a dilemma. He can take a day off work to mend the pipe, or alternatively, he can employ a professional plumber to repair the damaged pipe. If the home-owner repairs the pipe, then a day's fees will be lost; it is a much cheaper option to employ a plumber. Realizing this, the homeowner opts for the financially expedient solution and employs the plumber. As a result both parties benefit from the transaction; the plumber gets paid a fee and the lawyer is able to earn more money in court. This analogy is very similar to trading greenhouse gas emission permits, insomuch as those parties buying emission *permits* or *credits* are actually paying someone else to reduce greenhouse gas emissions who can

Table 1.4   Global energy consumption by fuel type for 1999 (compiled from BP energy data) [2]

Primary energy: consumption by fuel* (Million tonnes oil equivalent)	1999 Oil	1999 Natural gas	1999 Coal	1999 Nuclear energy	1999 Hydro-electric	1999 Total	1999 Per cent of total(%)
USA	888.9	561.7	546.4	197.8	27.2	2222.0	25.92
Canada	82.9	65.4	27.8	19.0	29.6	224.6	2.62
Mexico	80.8	30.6	6.0	2.6	2.8	122.8	1.43
Total North America	1052.6	657.7	580.2	219.4	59.6	2569.4	29.97
Argentina	20.8	30.4	0.7	1.8	1.9	55.6	0.65
Brazil	83.2	6.4	11.7	1.0	25.2	127.5	1.49
Chile	11.7	3.6	4.3	–	1.2	20.7	0.24
Colombia	10.6	4.7	2.1	–	2.9	20.3	0.24
Ecuador	6.0	0.1	–	–	0.6	6.7	0.08
Peru	7.4	0.5	0.5	–	1.2	9.5	0.11
Venezuela	21.9	24.1	0.3	–	5.2	51.5	0.60
Other South and Central America	55.1	9.3	0.5	–	6.7	71.6	0.84
Total South and Central America	216.7	79.1	20.1	2.8	44.9	363.4	4.24
Austria	12.1	7.2	3.2	–	3.6	26.1	0.30
Belgium and Luxembourg	32.4	13.3	6.9	12.7	0.2	65.4	0.76
Bulgaria	4.9	2.8	6.6	3.9	0.2	18.6	0.22
Czech Republic	8.2	7.7	17.1	3.4	0.2	36.7	0.43
Denmark	10.6	4.5	4.7	–	†	19.9	0.23
Finland	10.7	3.3	3.6	6.0	1.1	24.6	0.29
France	96.4	33.9	14.2	101.9	6.6	253.0	2.95
Germany	132.4	72.1	80.2	43.9	2.0	330.6	3.86
Greece	18.7	1.2	8.8	–	0.4	29.1	0.34
Hungary	7.1	9.9	3.4	3.6	†	24.1	0.28
Iceland	0.9	–	0.1	–	0.5	1.5	0.02
Republic of Ireland	8.3	3.0	1.6	–	0.1	13.1	0.15
Italy	94.4	56.0	11.6	–	3.9	165.9	1.94
Netherlands	40.6	34.1	7.0	1.0	†	82.7	0.96
Norway	10.1	3.2	0.7	–	10.5	24.4	0.28
Poland	19.9	9.3	61.0	–	0.4	90.5	1.06
Portugal	15.4	2.0	3.6	–	1.1	22.2	0.26
Romania	9.5	15.5	6.7	1.3	1.6	34.5	0.40
Slovakia	3.0	5.8	4.8	3.4	0.4	17.4	0.20
Spain	68.4	13.5	20.5	15.2	2.6	120.2	1.40
Sweden	16.1	0.8	2.0	18.9	6.2	43.9	0.51
Switzerland	12.6	2.4	0.1	6.4	3.5	25.1	0.29
Turkey	29.5	10.8	27.7	–	3.0	71.0	0.83
United Kingdom	79.4	82.8	35.6	24.8	0.7	223.4	2.61
Other Europe	16.3	4.6	10.1	1.2	2.7	35.0	0.41
Total Europe	757.9	399.7	341.8	247.6	51.5	1798.9	20.99
Azerbaijan	6.3	5.0	–	–	0.1	11.5	0.13

Belarus	6.1	13.8	0.1	–	†	20.0	0.23
Kazakhstan	6.0	7.1	19.8	–	0.6	33.6	0.39
Lithuania	3.1	2.2	0.1	2.5	0.1	8.0	0.09
Russian Federation	126.2	326.4	109.4	30.9	13.8	606.8	7.08
Turkmenistan	4.5	10.2	–	–	–	14.7	0.17
Ukraine	13.3	63.6	38.5	18.6	1.0	135.0	1.57
Uzbekistan	7.1	44.3	1.8	–	0.6	53.8	0.63
Other Former Soviet Union	4.7	7.1	1.0	0.6	3.1	16.4	0.19
Total former Soviet Union	177.3	479.7	170.7	52.6	19.3	899.8	10.50
Iran	58.4	49.5	1.0	–	0.4	109.3	1.28
Kuwait	8.6	7.8	–	–	–	16.4	0.19
Qatar	1.1	14.3	–	–	–	15.4	0.18
Saudi Arabia	60.9	41.6	–	–	–	102.5	1.20
United Arab Emirates	13.0	28.3	–	–	–	41.3	0.48
Other Middle East	65.0	19.1	5.7	–	0.3	90.1	1.05
Total Middle East	207.0	160.6	6.7	–	0.7	375	4.37
Algeria	8.1	20.0	0.3	–	†	28.4	0.33
Egypt	27.8	12.9	0.9	–	1.1	42.7	0.50
South Africa	21.8	–	82.1	3.5	0.3	107.7	1.26
Other Africa	58.2	14.0	6.8	–	4.9	83.9	0.98
Total Africa	115.9	46.9	90.1	3.5	6.3	262.7	3.06
Australia	38.0	17.8	45.5	–	1.5	102.8	1.20
Bangladesh	3.2	7.5	0.2	–	0.1	10.8	0.13
China	207.2	19.3	512.7	3.8	16.8	759.7	8.86
China Hong Kong SAR	9.3	2.4	3.9	–	–	15.6	0.18
India	95.2	21.4	154.5	3.3	7.1	281.5	3.28
Indonesia	46.8	24.8	10.5	–	0.8	82.9	0.97
Japan	257.3	67.1	91.5	82.0	8.0	505.9	5.90
Malaysia	20.3	17.1	1.2	–	0.4	39.0	0.45
New Zealand	6.3	4.7	1.2	–	2.0	14.1	0.16
Pakistan	18.2	15.6	2.1	†	1.8	37.7	0.44
Philippines	18	†	2.9	–	0.7	21.6	0.25
Singapore	28.3	1.4	–	–	–	29.6	0.35
South Korea	99.7	16.8	38.2	26.6	0.5	181.9	2.12
Taiwan	39.9	5.6	24.9	9.9	0.8	81.1	0.95
Thailand	35.4	15.6	7.9	–	0.3	59.2	0.69
Other Asia Pacific	18.6	4.4	53.1	–	3.5	79.6	0.93
Total Asia Pacific	941.7	241.5	950.3	125.6	44.3	2303	26.87
Total world	3469.1	2065.2	2159.9	651.5	226.6	8572.2	100.00
Of which: OECD	2178.1	1135.1	1070.2	565.8	118.2	5067.5	59.12
European Union 15	635.9	327.7	203.5	224.4	28.5	1420.1	16.57
Other EMEs‡	1079.8	421.7	890.8	23.3	84.2	2499.4	29.16

* In this Table, primary energy comprises commercially traded fuels only.

† Less than 0.05.

‡ Excludes Central Europe and Former Soviet Union.

do it more cheaply than they can themselves. However, in order for a trading scheme to work there must be something forcing the participants to make emission reductions, in much the same way that the home-owner was forced to act because the burst pipe would have created a considerable amount of damage if left unattended. Therefore, at the heart of any trading scheme there must be an obligation on the part of the participants to achieve greenhouse gas emission reductions; the obligations under the Kyoto Protocol are forcing countries to reduce greenhouse gas emissions. However, it does not matter geographically where these reductions are made. It is, therefore, possible for participating companies to trade all or part of their obligation without any detrimental environmental effects. This can be achieved by issuing permits which allow the holder to emit a given quantity of greenhouse gas.

The rules for any emissions trading system have yet to be fully worked out and there are many difficult international problems which must be solved, not least technical issues associated with the administration, verification and enforcement of such a scheme. One possible way in which such a scheme might work would be to allow permit holders to sell some of their 'excess capacity'. For example, in 1990 greenhouse gas emissions from many FSU countries were high. However since the break-up of the Soviet Union there has been a severe decline in economic activity in this region with the result that by 2000 emission levels were much lower than those in 1990. Under the Kyoto Protocol Russia and Ukraine are required to stabilize greenhouse gas emissions at a 100% of their 1990 emission levels. It is predicted that by 2010 in these countries emissions levels will still be below 1990 levels. Consequently, these countries would have a considerable block of surplus permits to sell. Western European countries might find it cost effective to purchase some of the 'excess capacity' in the FSU countries rather than reduce their own CO_2 emissions. Alternatively, countries with high labour and infrastructure costs could invest in emission reducing activities in countries with lower costs and claim emission reduction credits for themselves, thus offsetting their 'overseas' activities against their own emission reduction targets.

One possible advantage of tradable emissions permits is that they can be rationed. Permits could be issued which would give the parties involved in the scheme the right to emit at their 1990 emissions levels for free. If every year a proportion of permits were to be withdrawn and then sold either by direct sale or by auction, incentives would be created for the participants in the scheme to reduce emissions. Also, if permits were tradable and banking was to be permitted then a market would be created. However, it is essential that a single, central authority be created which could control the system and fine parties who exceeded their quotas.

1.6 Energy consumption

Global energy consumption data for 1999 is presented in Table 1.4 (see pages 16–17). These data give a detailed breakdown of energy consumption on a country-by-country and fuel-by-fuel basis. It can be seen from Table 1.4 that in 1999 global energy consumption was approximately 8572 mtoe. Of this total approximately 25.9% was consumed in the USA, 16.6% in the EU and 10.5% in the FSU countries. Indeed, 59.1% of all global energy consumption occurred in the OECD countries. This demonstrates that currently the major part of global energy consumption takes place in industrialized countries.

The fuel mix and hence the carbon intensity varies greatly from region to region. In the USA, for example, the ratio of coal to natural gas consumption is 1:1.03, whereas in the UK the ratio is 1:2.33, while in China the ratio is only 1:0.04 and in India it is 1:0.14. These figures reflect the fact that, firstly in the UK during the 1990s there has been a trend away from coal and towards natural gas, and secondly there is a heavy reliance on coal in China and India.

The Energy Information Administration (EIA) in the USA predicts that in the twenty-first century there will be substantial increases in energy demand, based mostly on fossil fuels [3]. This is expected to occur mainly because of economic growth in the developing economies of Asia and South America. In developing countries, energy and economic growth tend to move in parallel. Economic development is an energy intensive process which ultimately raises living standards and facilitates broad access to electricity and motorized transportation. Economic development of the infrastructure also causes growth in energy intensive manufacturing industries. In contrast to the developing countries, in advanced industrialized countries the link between economic growth and energy consumption is relatively weak, with energy demand growth lagging behind economic growth. In advanced economies per capita energy use tends to be relatively stable, with old energy intensive appliances and equipment often being replaced by newer more energy efficient equipment. Consequently it is predicted that the percentage growth in energy demand will be considerably lower in the developed countries compared with the developing countries.

It is predicted by the EIA that the trend towards the increased use of natural gas in the 1990s will continue in the twenty-first century [3]. This is because natural gas is increasingly perceived as the fuel of choice for electricity generation: it has a much lower carbon intensity than coal; the electricity generation process is more efficient; it is free from the industrial disputes which are often associated with coal production; and it is much cheaper and quicker to construct gas-fired power plants. Oil demand is predicted to grow with the increased use of motorized transport in developing countries.

1.7 Energy reserves

One of the major concerns of environmentalists and economists alike is the rate at which 'precious' fossil fuel reserves are being expended. It has been estimated that crude oil could remain plentiful and cheap for at least 40 years [24]. This prediction 'assumes that production volumes will remain constant, and assumes that production rates can be maintained as reserves decline' [24]; this is an assumption which appears optimistic given that the demand for oil should increase with economic growth in the developing world. The EIA states:

> Worldwide, oil remains the dominant source of energy throughout the projection horizon [i.e. to 2020], as it has since 1970. Oil's key role [is] in the transportation sector – where it does not currently have any serious competition from other energy sources...[3]

Even if oil remains plentiful until the middle of the twenty-first century, geophysical surveys indicate that massive new oil deposits are unlikely to be found [24]. In future new oil finds are likely to be much smaller than those already discovered. In other words, the law of diminishing returns is likely to apply. Consequently there will be a pressing need to find alternative sources of fuel to replace oil as the choice of fuel for motorized transportation.

Although there is considerable debate about the global oil reserves, it is generally agreed that conventional oil production outside the Middle East will start to decline before that of the Middle East, implying a greater reliance on Middle Eastern oil [2]. Table 1.5 shows global proven energy reserves at the end of 1999. From these data it can be seen that approximately 65% of proven oil reserves are located in the Middle East. It is, therefore, not surprising that as existing supplies become depleted the rest of the world will become reliant on Middle Eastern oil suppliers.

Table 1.5 Proven energy reserves by fuel type (compiled from BP energy data) [2]

Proven reserves at the end of 1999	Oil		Natural gas		Coal	
	Thousand million barrels	Share of total(%)	Trillion cubic metres	Share of total(%)	Million tonnes	Share of total(%)
Total North America	63.9	6.18	7.31	5	256477	26.10
Total South and Central America	89.5	8.66	6.31	4.30	21574	2.20
Total Europe	20.7	2.00	5.15	3.50	122032	12.40
Total Former Soviet Union	65.4	6.32	56.7	38.70	230178	23.40
Total Middle East	675.7	65.35	49.52	33.80	n a	n a
Total Africa	74.8	7.23	11.16	7.70	61605*	6.20*
Total Asia Pacific	44.0	4.26	10.28	7.00	292345	29.70
Total World	1034.0	100.00	146.43	100.00	922606	100.00

* Includes the Middle East.

It is likely that as global oil stocks become depleted natural gas will replace oil, since gas can be converted into liquid fuels which can be used for automotive purposes at relatively low cost. The EIA expects natural gas to be the fastest-growing primary energy source between 1996 and 2020 [3].

References

1. Campbell, D. (2001). Blackouts bring gloom to California. *The Guardian*, 19th January.
2. BP energy statistics, 2001, BP Website (www.bp.com/downloads/702/BPwebglobal.pdf) (March 2002).
3. International Energy Outlook 1999 (IEO99) (1999). World Energy Consumption. DOE/EIA-0484 (99).
4. Digest of United Kingdom Energy Statistics 2000 (2000). DTI.
5. Hill, R., O'Keefe, P. and Snape, C. (1995). *The future of energy use*. (Chapter 1), Earthscan.
6. World Development Report for 1990 produced by the World Bank.
7. International Energy Outlook 1999 (IEO99) (1999). Environmental Issues and World Energy Use. DOE/EIA-0484 (99).
8. WMO Annual Statement on the Global Climate, 1998, WHO Website (www.wmo.ch/web/Press/Press626.html) (March 2002).
9. NASA Goddard Institute for Space Studies, Global Temperature Trends 1998 Summation, NASA Website (www.giss.nasa.gov/research/observe/surftemp/1998.html) (March 2002).
10. Climate change: The IPCC scientific assessment (1990). WMO/UNP Intergovernmental panel on climate change.
11. May, R. Global Warming Needs Action (1998). *Energy Exploration and Exploitation*, **16**(1), 93–102.
12. Department of the Environment, Transport and the Regions (2000). Climate Change: Action to Tackle Global Warming.
13. Shorrock, L. D. and Henderson, G. (1990). Energy use in Buildings and Carbon Dioxide Emissions. BRE Report.

14. Beggs, C. B. (1996). A method for estimating the time-of-day carbon dioxide emissions per kWh of delivered electrical energy in England and Wales. *Building Services Engineering Research and Technology*, **17**(3), 127–34.
15. Energy Information Administration, DOE, USA. September 2001, EIA Website (www.eia.doe.gov/emeu/cabs/uk.html) (March 2002).
16. Harrison, R. M. (1990). *Pollution: causes, effects and control* (Chapter 9). Royal Society of Chemistry.
17. Environmental Indicators: Ozone Depletion (1996). United States Environmental Protection Agency.
18. The Accelerated Phase out of Class I Ozone-Depleting Substances (1999). United States Environmental Protection Agency.
19. HCFC Phase out Schedule (1998). United States Environmental Protection Agency.
20. Energy in the 21st-century: The return of geopolitics? (1999). Highlight of the OECD Information Base, OECD, No. 17, May.
21. Greenhouse gases and climate change – Environment in EU at the turn of the century (Chapter 3.1) (2001). European Environment Agency.
22. Svendsen, G. T. (1998). Towards a CO_2 Market in the EU: the case of electric utilities. *European Environment*, **8**(4), 121–8.
23. Kopp, R. and Toman, M. (2001). International Emissions Trading: A Primer: www.weathervane.rff.org/features/feature049.html. (12 July).
24. Campbell, C. J., Laherrère, J. H., George, R. L., Anderson, R. N. and Fouda, S. A. (1998). Preventing the next oil crunch. *Scientific American*, **278**(3), 59–77.

Bibliography

Climate change: The IPCC scientific assessment (1990). WMO/UNP Intergovernmental panel on climate change.

Harrison, R. M. (1990). *Pollution: Causes, effects and control.* (Chapter 9) Royal Society of Chemistry.

Hill, R., O'Keefe, P. and Snape, C. (1995). *The future of energy use.* (Chapter 1) Earthscan.

International Energy Outlook 1999 (IEO99) (1999). Environmental Issues and World Energy Use. DOE/EIA-0484 (99).

International Energy Outlook 1999 (IEO99) (1999). World Energy Consumption. DOE/EIA-0484 (99).

2

Utility companies and energy supply

Energy costs are strongly influenced by the policies of utility companies and fuel suppliers. It is therefore important to have a good understanding of the supply side of the energy industry. This chapter explains the fundamental issues associated with the supply of fuel and energy, and explains how utility companies recover their costs.

2.1 Introduction

This chapter is not intended to be a text on the chemical properties of various fuels, or indeed a work on the geographical aspects of energy supply, rather it is written from the point of view of the end user, who may also be the energy purchaser. The aim of the chapter is to explain the ways in which utility companies operate and also to highlight key issues which influence the supply of fuel and energy.

When investigating the subject of energy supply it is necessary to understand the process by which energy is delivered to the consumer. In general terms, most facilities consume one or more of the following 'fuels':

- Coal
- Oil
- Natural gas
- Electricity

Natural gas and electricity are delivered to consumers by utility companies via extensive pipe and cable infrastructures. By contrast, coal and oil are purchased on the open market and require vehicular delivery and storage facilities. Each 'fuel' therefore has its own set of peculiarities and limitations, and these strongly influence both usage and energy costs. Some fuels, such as natural gas, are unavailable in many locations because nearby pipelines do not exist. Other fuels, such as coal and oil, are perceived as dirty and difficult to handle, and also require considerable storage space. Since all these factors influence energy costs, it is important to be familiar with the costs associated with each type of fuel. Fuel and energy costs can be categorized as:

- Direct costs associated with the purchase of the 'fuel';
- Indirect cost associated with the use of the 'fuel'; and
- Indirect environmental costs.

Direct fuel costs are those costs specifically associated with the purchase of the fuel, such as unit energy charges and standing charges, and indirect costs are those costs associated with the storage and handling of fuels such as coal and oil. Indirect environmental costs also exist, but the end user does not always perceive their existence. Environmental costs can take an obvious form, such as an environmental tax on fossil fuel consumption (see Chapter 1). However, it is more often the case that environmental costs are absorbed and disguised within other costs. For example, if an electricity company is required to invest in expensive desulphurization equipment to clean up its generating plant, it will probably pass this environmental related cost on to its customers in the form of higher energy prices.

2.2 Primary energy

Three fuels, oil, coal and natural gas dominate the world's primary energy market. In 1999, 40.6% of the world's primary energy came from oil, 25.0% from coal, 24.2% from natural gas, 2.7% from hydro-electricity and 7.6% from nuclear power [1]. Although oil is currently the principal source of the world's energy, consumption is outstripping supply and known oil reserves are dwindling. It has been estimated that at current rates of consumption, cheap available oil supplies will be exhausted in approximately 40 years time [2]. However, unlike oil, the Earth's coal reserves are in a much healthier state, with worldwide reserves of anthracite and bituminous coal alone exceeding 200 times annual coal consumption [1]. One might therefore assume that coal could fill the 'vacuum' which will be created as oil reserves diminish. This however is a simplistic assumption, since oil is primarily used in the transport sector and coal, unless synthesized, is unsuitable for this purpose. Also, coal is perceived as being environmentally unfriendly, producing large quantities of carbon dioxide (CO_2) and sulphur dioxide (SO_2) when burnt. By contrast natural gas is clean when burnt and is relatively easily to handle. Consequently, worldwide natural gas consumption is increasing. For example, in the UK, natural gas consumption rose from 52.4 billion cubic metres in 1990 to 92.0 billion cubic metres in 1999 [1]. The expansion in the natural gas sector in the UK was mainly due to the construction of a large number of new combined-cycle gas turbine power stations in the 1990s. The rise in the UK's gas consumption was matched by a steep decline in its coal consumption, which was only 35.6 mtoe in 1999, compared with 64.9 mtoe in 1990 [1]. This scenario has been matched elsewhere in Europe. In France, for example, coal production fell by 59.0% from 1990 to 1999 [1] and in Germany the corresponding fall was 51.3% [1].

In terms of global energy consumption, natural gas is the fastest growing primary fuel. This is primarily because natural gas reserves are large and the cost of production is relatively low. In the FSU countries alone, proven gas reserves are 56.7 trillion cubic metres [1999 data], enough to supply their own, together with the whole of Europe's, requirements for at least 50 years [1]. One important attribute of natural gas is that it is very clean compared with more carbon intensive fuels such as oil and coal. Following the 1997 Kyoto agreement many nations have sought to reduce the CO_2 emissions by increased utilization of natural gas. It is also a more manageable fuel than coal, which has to be mined and involves expensive handling. Consequently, in many parts of the world electricity companies are turning away from coal and switching to natural gas. One added bonus to the generating companies is that gas fired power stations are not prone to industrial action by miners, as coal fired stations are.

2.3 Delivered energy

It is important to appreciate the difference between primary energy and delivered energy. For example, natural gas is a primary fuel which can be burnt in a power station to produce electricity (i.e. a 'secondary' fuel) or instead piped straight to the consumer as a primary fuel. By contrast, electricity is always a 'secondary' fuel, which is produced from a primary source, often by a very inefficient conversion process. Therefore for every unit (i.e. kWh) of electricity which is delivered to a property, several units of primary energy must be consumed.

The price paid by the consumer for energy usually reflects closely the cost of its production and its general availability. Consider the case of electricity produced from natural gas in a combined-cycle power station, with thermal efficiency of 47%. It is not difficult to see that the cost to the utility company of producing the electricity must be over twice the cost of the natural gas. Indeed, when all the other maintenance, transportation and management costs are factored in, then the cost of the delivered electricity will much higher. In fact, in the UK the unit price of electricity is generally three to five times the price of natural gas, depending on the type of consumer.

2.4 Electricity supply

Consider the generic model of an electricity supply network shown in Figure 2.1. The saleable commodity, electricity, is generated in power stations, from a variety of primary energy sources. These power stations are all connected to a high voltage transmission grid, which is used to transmit the electricity over long distances. At various points in the system, electricity is drawn from the transmission grid and distributed to consumers. This involves the use of a local distribution network which distributes the electricity at a reduced voltage, say 33 kV or 11 kV, to local sub-stations, where the voltage is further reduced, to say 415 V, before being supplied to the consumer. Finally, the electricity is sold to consumers and the amount consumed recorded using meters.

In broad terms, the various component parts of an electric supply industry can be categorized as follows:

(i) The generation process: This takes place in the power stations and involves the conversion of primary energy from fossil fuels into electricity. It is at the generating stage that all the

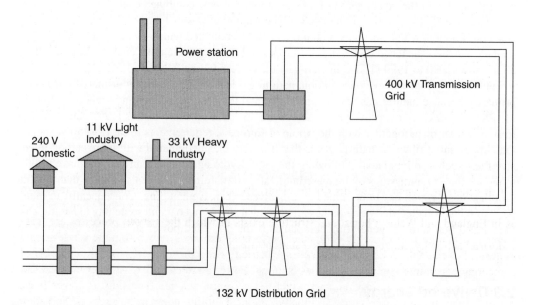

Figure 2.1   A generic electricity supply network

pollution associated with electricity production is created. Electricity generation is a complex and costly business, which involves the construction, operation and maintenance of large power stations and the purchase, transportation and storage of primary fuels. Because electricity cannot be stored easily it is necessary for the generating companies to have enough spare capacity to cope with the high peaks in demand which occur at certain times of the year. Therefore, many smaller and less efficient power stations are rarely used and only operate when demand on the transmission grid is high. This situation is very uneconomic because even though these power stations are infrequently used, they still need to be maintained. Consequently, the generating companies have to recover the cost of maintaining inactive power stations from the electricity produced by the active ones.

(ii)   The transmission process: This is the process whereby electricity is transported through a transmission grid over long distances around a region or country. The transmission grid is operated at a very high voltage, for example, 400 kV, in order to minimize the energy wastage. Operating a transmission grid involves the construction and maintenance of a very large piece of infrastructure, which may extend for thousands of miles, sometimes over very inhospitable terrain. Naturally large costs are involved in operating such a network and these must be recovered from the sale of any electricity which is transmitted through the grid.

(iii) The distribution process: Once electricity has been transmitted over a long distance at high voltage, it must then be distributed to the various end users. At various points along the transmission grid electricity will be 'siphoned' off into local distribution networks. These distribution networks are regional grids, operating at a lower voltage (e.g. 132 kV and 33 kV), which distribute the electricity around a city or a particular locality. During the distribution process the voltage of the electricity is stepped down (through the use of sub-stations) to the voltage required by the consumers (e.g. 240 V, 415 V or 11 kV). As with the transmission grid, the costs involved in maintaining and operating a distribution network must be recouped from the revenue received from the electricity sold.

(iv)   The sales process: The sales process is not as easy to identify as the other processes because it does not involve any obvious hardware, but it is no less important. All utility companies have to market their product in order to attract customers, and once customers have been found a utility company must monitor and record all the energy that is consumed, so that customers can be billed and the revenue collected. There are therefore considerable administration costs associated in managing each customer's account. These costs are usually recovered by levying a periodic standing charge on each customer.

The relationship between the various processes in a typical electricity supply industry is shown in Figure 2.2.

Although the processes described above are common to all electricity supply industries, the way in which they are achieved in practice varies considerably around the globe. In some countries all the four processes are performed by a single vertically integrated utility company, which generates, transmits and distributes electricity, and bills its own customers. Examples of vertically integrated utility companies exist in many parts of the USA. In other countries such as in England and Wales a horizontal structure exists in which the various processes are frag-mented, with a number of utility companies performing different roles within the whole supply industry. For example in England and Wales, electricity generation is performed by a number of competing generating companies such as National Power and PowerGen. Another independent company, the National Grid Company which levies rental charges for using its power lines, performs the transmission process. Finally, regional utility companies such as Yorkshire Electricity and London Electricity purchase electricity from the national grid and distribute it to

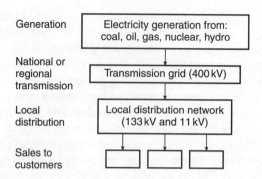

Figure 2.2 Typical structure of an electricity supply network

their customers. They also read their customers' meters and invoice them for the electricity consumed. This horizontal approach can result in extremely complex trading mechanisms (see Chapter 3). No matter how complex the structure, it is important to understand that the component processes described above and their related costs are the same the world over. Consequently the tariffs levied by all electricity utility companies tend to follow a similar format.

2.4.1 Electricity charges

From the discussion in Section 2.4 it can be seen that many costs are incurred in generating and supplying electricity, and these must be recovered from the end user. They can be summarized as:

- The cost of purchasing the primary energy and converting it to electricity;
- The cost of transporting electricity around a region or country;
- The cost of distributing electricity to the customer; and
- The cost involved in meter reading, billing and managing customers' accounts.

Because electricity cannot be stored, the size, and hence the cost, of the supporting infrastructure is governed by the maximum instantaneous load on the system and not the amount of energy which is consumed. Consider the electricity demand profiles for 1993 for the national grid in England and Wales presented in Figure 2.3. It can be seen from these profiles that the greatest demand for electricity in 1993 occurred in winter, reaching a peak of 48 GW at approximately 18.00 hours on 29th November [3]. However, for most of the year the demand was considerably less. In fact at 06.00 hours on the 1 August 1993 demand on the grid dropped to a low of approximately 16 GW. The only conclusion which can be drawn from this is that for most of the year the transmission and distribution grids are working at a level well below their maximum capacity. Although a considerable amount of generating capacity is idle for much of the year, it still has to be maintained on stand-by, ready to be used should demand on the grid rise to a high level. In fact, if excess capacity is not built into the system, when demand for electricity rises to a higher level, one of two things will happen:

1 There may not be enough generating capacity to meet demand; or
2 The cables in the transmission grid may become overloaded.

If either of these happen, then power cuts will occur, which is highly undesirable. Given this, the utility companies are faced with a decision, either to:

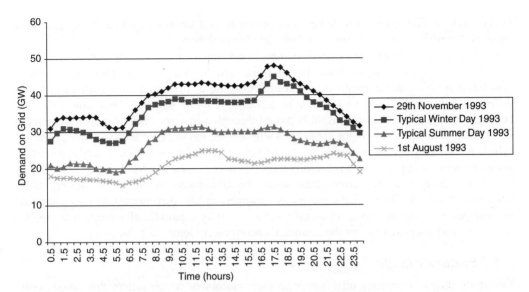

Figure 2.3   Electricity demand profiles for England and Wales for 1993[3]

- Build more power stations and reinforce their transmission and distribution networks, which is a very costly solution; or
- Discourage their customers from consuming large amounts of electricity at times when electricity demand is high.

Given this choice it is not surprising that most electricity utility companies opt for the latter solution. In practice they discourage high electricity usage by levying large demand charges on customers who use electricity during periods of peak demand. In northern Europe the period of peak demand is in the winter months, but in the southern states of the USA the peak occurs in summer due to the increased use of air conditioning equipment [4]. From this it can be seen that utility companies can influence demand on their network by altering the charges levied on customers under tariffs.

Although tariffs can be used to discourage customers, they can also be used to encourage them. Figure 2.3 demonstrates that electrical demand is much higher in the day time than at nighttime. Consequently, generating plant is under-utilized at night. In an attempt to remedy this situation, utility companies often offer nighttime (i.e. off-peak) electricity at a discounted rate as an incentive to customers to use off-peak electricity. This is a common practice worldwide. Some utility companies even divide the day into 3 or 4 bands and charge various unit energy rates for each band, depending on the demand experienced on their network. Utility companies also use tariffs to discourage building designers from over-designing their electrical installations by levying availability charges. When a building is constructed the local electrical utility company is generally obliged to supply it with electricity. This usually involves laying a new power cable to the building and may involve the construction of a new transformer and sub-station. In some cases, if the facility is large, it might even involve reinforcing the whole local electrical distribution network. The capital cost involved in all this work is borne by the utility company. Not surprisingly the utility company does not look kindly on end users who over-estimate their power requirements as this results in them incurring large capital costs. One solution is therefore to levy a supply availability charge on the power demand (i.e. system rated capacity) requested by the end user. This charge is levied each month on every 1 kW (or 1 kVA) of the system rated capacity. Clearly a deterrent to over-cautious building designers!

Some utility companies meter their customers in kW and kWh rather than kVA and kVAh. At school we are taught that electrical power (measured in Watts) is the product of voltage (measured in Volts) and current (measured in Amps). So at first sight there appears to be no difference between kW and kVA. There is, however, a subtle and very important difference between the two. If an electric current is passed through a reactive load, such as a fluorescent lamp fitting or an induction motor, the current will become out of phase with the voltage and lag behind it. This subject is discussed at length in Chapter 12, so here it suffices to say that reactive loads consume more power than is usefully used. Therefore, a reactive load such as an induction motor will draw a larger current than would be anticipated by its useful power rating. Consequently, if a utility company meters its customers in kW, then it must levy an additional reactive power charge in kVArh. This ensures that the utility is paid for all the power it supplies to the site, and not penalized for the customer's poor power factor.

2.4.2 Electricity tariffs

Electricity utility companies offer tariffs to their customers which reflect the various costs incurred. These tariffs are published in advance and are standard for all customers. Although tariffs differ from one utility company to another, they are all structured to recover the various costs incurred in generating, transmitting, distributing and selling electricity.

Domestic tariffs are probably the simplest form of electricity tariff that exists. They are offered to domestic and other small consumers, and are usually billed quarterly. In their simplest form they consist of a fixed standing charge levied every quarter and on a standard unit charge, as shown in Table 2.1. A more sophisticated version of the simple domestic tariff is the introduction of peak and off-peak unit charges, designed to encourage off-peak consumption of electricity. In the UK this tariff is known as an 'economy seven tariff', a typical example of which is shown in Table 2.1.

For larger non-domestic customers with demands below 50 kW, 'block tariffs' are usually offered. Block tariffs are similar in structure to domestic tariffs with the exception that the first tranche of energy consumed, usually a 'block' of approximately 1000 kWh, is charged at a higher unit rate. Some utility companies may even offer tariffs with more than one block. As with the domestic tariff, peak/off-peak variants are usually offered. Table 2.2 shows the structure of a typical block tariff.

For larger commercial and industrial consumers the most commonly offered type of tariff is probably the 'maximum demand tariff', which is offered in various guises by many utility companies around the world. Table 2.3 shows a typical 'maximum demand' tariff used by a UK electricity utility company.

Table 2.1   Domestic electricity tariffs

General Domestic Tariff	
Quarterly standing charge	£ 8.25
Unit charge	6.53 p/kWh
Economy Seven Domestic Tariff	
Quarterly standing charge	£ 10.78
Off-peak unit charge (00.30–07.30)	2.28 p/kWh
Peak unit charge (all other times)	6.89 p/kWh

Table 2.2   Block electricity tariff

General Block Tariff	
Quarterly standing charge	£ 8.69
Unit charge for the first 1000 kWh consumed	8.94 p/kWh
Unit charge for additional kWh consumed	6.70 p/kWh
Economy Seven Block Tariff	
Quarterly standing charge	£ 11.22
Off-peak unit charge (00.30–07.30)	2.28 p/kWh
Peak unit charge for the first 1000 kWh consumed	8.94 p/kWh
Peak unit charge for additional kWh consumed	7.05 p/kWh

Under a maximum demand tariff shown in Table 2.3 the billing and administrative costs are recovered through a monthly standing charge which is constant and independent of the amount of electricity consumed. The unit charge covers the cost of providing electrical energy and therefore is levied on every kVAh of electricity consumed. The unit charge is generally made up of two components: a larger component, which covers the cost of producing the electricity, and a smaller component, which covers the operating costs associated with its transmission and distribution. In order to encourage off-peak electricity consumption, a reduced off-peak unit charge is offered nightly from 00.30 to 07.30 hours. This encourages customers to load shift and thus reduce the demand on the network during periods of peak demand.

The capital costs associated with the distribution network are recovered through the maximum demand charge, which is levied to penalize consumers who have a high power demand (in kW or kVA) during periods in which demand for electricity is high. In the case of the tariff shown in Table 2.3, the 'peak power demand' period lasts from 1 November to 28 February. During this period a demand charge is levied on every kVA of the peak instantaneous demand recorded in any particular month. During the colder months of December and January the charge is £5.40 per kVA. However, this drops to £2.70 per kVA during November and February. For the rest of the year, no demand charges are levied.

The supply availability charge is designed to deter customers and building designers from requesting oversized supply cables to their facilities. If a customer over-estimates their required

Table 2.3   Typical maximum demand tariff

	Tariff
1. Monthly charge	£32.00
2. Supply availability charge per month for each kVA of chargeable supply capacity	£1.40
3. Maximum demand charge per month for each kVA of monthly maximum demand in the months of:	
December and January	£5.40
November and February	£2.70
4. Unit charges for maximum demand tariffs (p/kVAh)	
For each unit supplied	6.15 p
For each unit supplied between the hours of 00.30 and 07.30 each night	2.65 p

Table 2.4   Typical Seasonal Time of Day Tariff

	Tariff
1. Monthly charge	£32.00
2. Supply availability charge per month for each kVA of chargeable supply capacity	£1.40
3. Winter unit charges (p/kVAh) (November to March inclusive)	
(i) For each unit supplied between 16.00 and 19.00 each day on Mondays to Fridays inclusive during	
(a) December and January	30.50 p
(b) November and February	17.60 p
(ii) For each unit supplied between 08.30 and 20.00 each day on Mondays to Fridays inclusive during the months of December to February inclusive, other than charged at rates (i) (a) or (b) above	7.45 p
(iii) For other winter day units supplied between 07.30 and 00.30	5.30 p
(iv) For winter night units supplied between 00.30 and 07.30	2.65 p
4. Summer unit charges (p/kVAh) (April to October inclusive)	
(i) For each summer day units supplied between 07.30 and 00.30	4.50 p
(ii) For summer night units supplied between 00.30 and 07.30	1.90 p

supply capacity, then the utility company will incur additional expense. If the customer's high demand estimates do not materialize, then the utility company will have spent money which it cannot recoup through either maximum demand or unit charges. Consequently, an availability charge is levied monthly on each kVA or kW of the requested supply capacity, in order to penalize over-ambitious customers.

A variation on the maximum demand tariff described above is the seasonal time of day (STOD) tariff which is illustrated in Table 2.4.

Under a seasonal time of day tariff, monthly standing and availability charges are levied as in the maximum demand tariff. However, there are no demand charges, and instead the unit charges reflect the demand on the utility company's network, penalizing very heavily those customers who consume electricity during periods of peak demand.

Variations on the tariffs described above can be found throughout the world. One novel variation occurs in Finland, where IVO, the countries largest supplier, has a triple band tariff. This reflects the nature of, and the production costs associated with, the various types of generating plant required to meet demand [5]. The base tariff is linked to hydro and nuclear power, the middle tariff is related to coal-based power production, and the peak tariff is based on oil-fired and gas turbine power production. In many hotter regions, such as in the southern states of the USA, peak demand occurs during the summer months due to the intensive use of air conditioning plant. Therefore utility companies in these regions levy maximum demand charges in the summer months and not during the winter.

It should be noted that in many countries some form of 'value added tax' (VAT) is levied on energy consumption. If this applies in a particular region or country, it is important that this is considered when estimating and analysing electrical energy bills.

The discussion above demonstrates that the overall price per kWh or kVAh paid by the consumer depends greatly on the end user's consumption profile. Indeed, in some circumstances it can be considerably greater than the unit energy charge quoted by a utility company. Example 2.1 illustrates the technique which should be used in order to accurately calculate electrical energy costs.

Example 2.1

The monthly electricity consumption data for a hotel building for a calendar year is as follows:

Month	Electricity peak units (kVAh)	Electricity off-peak units (kVAh)	Monthly maximum demand (kVA)
January	223 475	63 031	592
February	214 718	57 076	552
March	186 845	46 711	476
April	185 513	40 722	471
May	174 103	33 162	427
June	165 988	27 021	381
July	163 181	24 383	365
August	150 089	26 486	372
September	164 746	31 380	410
October	174 983	41 045	452
November	189 752	47 438	486
December	222 952	55 738	623
Totals	2 216 345	494 193	n.a.

Given that the declared maximum supply capacity of the hotel building is 650 kW and that the maximum demand tariff shown in Table 2.3 applies, determine the annual electricity bill, and hence an average price for each unit (kVAh) of electricity consumed (exclusive of VAT).

Solution

Elemental charge	Calculation	Cost (£)
Monthly standing charge	£ 32.00 × 12	384.00
Supply availability charge	650 kVA @ £ 1.40 per kVA × 12	10 920.00
Maximum demand charges		
December	623 kVA @ £ 5.40	3 364.20
January	592 kVA @ £ 5.40	3 196.80
November	486 kVA @ £ 2.70	1 312.20
February	552 kVA @ £ 2.70	1 490.40
Energy charges		
Peak rate	2 216 345 kVAh @ 6.15 p per kVAh	136 305.22
Off-peak	494 193 kVAh @ 2.65 p per kVAh	13 096.11
	Annual electricity cost	**170 068.93**

$$\text{Average unit price of electricity} = \frac{170\,068.93 \times 100}{(2\,216\,345 + 494\,193)} = 6.274 \text{ p/kVAh}$$

2.5 Natural gas

Gas utility companies are similar to electricity utility companies insomuch as they both supply energy directly to buildings via pipes or cables. Both types of utility company have therefore to construct, operate, and maintain large transmission and distribution networks. In addition, both types of utility company have to invoice customers and maintain customer accounts. However, there are some distinct and influential differences between the two 'fuels':

Table 2.5   Composition of natural gas

Constituent part	Percentage by volume (%)
Methane, CH_4	92.6
Ethane, C_2H_6	3.6
Propane, C_3H_8	0.8
Butane, C_4H_{10}	0.3
Nitrogen, N_2	2.6
Carbon dioxide, CO_2	0.1

From Eastop and Croft (1990) *Energy Efficiency for Engineers and Technologists*,
© Longman Group Ltd 1990, reprinted by permission of Pearson Education Ltd.

- Unlike electricity, natural gas is a primary fuel which must be removed from the ground by some means;
- Natural gas can be stored in large quantities, while electricity cannot; and
- Unlike electricity, which is required all year round, the market for natural gas is very seasonal and is weather dependent.

Natural gas consists almost entirely of methane (see Table 2.5) and has a net calorific value of approximately 38.6 MJ/m^3; its density is about 0.73 kg/m^3. As such it has a low energy density compared with oil. In fact, 1 m^3 of natural gas contains approximately one-thousandth of the energy of the same volume of crude oil. Consequently, transportation of natural gas is an expensive and difficult operation, involving substantial networks of large diameter pipelines. Gas transportation is therefore usually by high-pressure pipeline, although gas can be liquefied and then transported by ship. Both methods usually require large-scale storage near to consumers to balance out fluctuations in demand.

Figure 2.4 shows an example of a supply chain for gas produced from an offshore platform [7]. From the platform the gas is sent through an offshore pipeline to a landing terminal. From there, it is either sent through transmission pipelines to serve the domestic market, or transferred to another country via an international transmission pipeline. At some stage the gas will end up in a regional distribution pipeline before eventually being supplied to domestic users, large industrial users, or electricity generators.

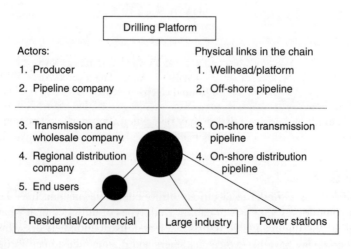

Figure 2.4   Gas chain[7]

Gas supply chains vary depending on their location and often involve a large number of parties. The gas producers may not necessarily be the companies who own the offshore pipeline; similarly the transmission company may not be involved in regional and final distribution. Therefore as gas flows along a chain, its ownership may be transferred several times.

2.5.1 Natural gas production, transmission and distribution

The gas chain in Figure 2.4 shows that a number of distinct processes are involved in bringing the finished product to the customer:

(i)   *Gas gathering pipelines*: These link the production fields to the transmission lines and may run undersea or over land. They transport untreated gas which is not usually of marketable quality. Untreated gas may be 'dry' or 'wet'. The gathering pipelines for 'dry gas', which is relatively free of liquids, will be much like any other transmission line. 'Wet gas' contains substantial quantities of liquid; if this is to be transported then it must be treated to prevent line clogging.

(ii) *Gas treatment*: When it is extracted, natural gas is nearly always mixed with impurities such as water, acid gases, nitrogen and helium [7]. In order to prevent corrosion of transmission lines, these impurities must be removed. This process usually takes place in an on-shore gas treatment facility.

(iii) *Gas transmission*: Gas transmission pipelines ensure an uninterrupted supply of gas at a set pressure from the gas gathering pipeline to various delivery points. These delivery points are typically the inlets to local utility distribution systems. At their simplest, transmission pipelines can simply be a connection between two points. However, it is more common that transmission pipelines supply several points along their routes, as is the case with the Russian pipelines, which serve several European countries. In transmission pipelines it is necessary to boost the gas pressure at regular intervals by using compressors.

(iv) *Blending*: Depending on their hydrocarbon and nitrogen content, the calorific content of natural gases from different sources may vary. Many gas appliances are designed to run on one type of gas and cannot burn a gas with a different calorific value. Consequently, blending stations are required to mix the various different gases into a blend whose calorific value is within the tolerance of end users' appliances.

2.5.2 Peak demand problems

In Europe and North America, where natural gas is predominantly used for space heating, demand is strongly dependent on outside air temperature. Gas transmission and distribution systems must therefore cope with large swings in seasonal demand, while still maintaining a continuous supply to customers. It is particularly important to ensure that all gas pipelines are maintained well above atmospheric pressure, so that air is not entrained into the system; otherwise a potentially explosive mixture may be created. In order to overcome the problem of fluctuating demand, gas utility companies can employ a number of alternative strategies:

- They can construct a transmission infrastructure which is large enough to cope with the maximum likely demand;
- They can store large reserves of gas near to consumers when gas demand is low, ready for withdrawal when demand is high;
- They can offer interruptible supply tariffs with discounted unit charges. However, under this type of tariff, the utility company does not guarantee to supply gas to customers when demand rises to a high level.

The first solution involves large capital expenditure on infrastructure; an infrastructure which is inevitably under-utilized for most of its life, with the result that production and transmission costs are increased. The second solution attempts to balance production and transmission loads and also provides a measure of security against supply disruption. However, it does involve capital expenditure on storage facilities. The third solution involves no capital expenditure and is very popular with utility companies. By offering an interruptible supply to large customers, the utility company is not guaranteeing a gas supply during periods of peak demand. Interruptible supplies are popular with organizations that are able to switch at short notice to another fuel, such as oil, and so are able to reduce the peak demand on the gas utilities network. In return, consumers who opt for an interruptible supply are offered an advantageous tariff. In practice, a combination of all three solutions is often employed to balance a utility company's supply load.

2.5.3 Gas tariffs

The supply of natural gas to customers involves costs which must be met by the various companies involved. These can be categorized as:

- The cost of collecting, processing and delivering the gas 'on-shore';
- The cost of transporting the gas through the transmission pipelines; and
- The costs involved in marketing, billing and managing customers' accounts.

One of the most influential factors in determining the price that customers pay for gas is the transportation levy that gas suppliers have to pay for the use of transmission pipelines. This transportation levy reflects the costs incurred in constructing and operating the transmission network, and is generally determined by the length of the pipeline and the maximum gas-flow rate during periods of peak demand.

Gas tariffs are much simpler than those used for electricity. They tend to follow a fixed pattern of charges, namely: a fixed quarterly standing charge and a charge for every unit (kWh) of gas supplied. Unit charges usually incorporate a transportation charge. For smaller consumers unit charges are generally constant. For larger customers, the unit charge usually decreases as gas consumption increases. Table 2.6 shows a typical gas tariff for a small customer.

It should be noted that in many countries VAT is levied on gas consumption. A variation on this gas tariff shown in Table 2.6 is the introduction of a higher unit charge for the first 'block' of gas consumed, in a similar manner to an electrical block tariff.

Larger customers tend to be offered monthly tariffs in which the unit charge decreases as gas consumption increases. They also have the option of negotiating either firm or interruptible supply contracts. The unit charges associated with an interruptible supply are lower than those for a firm supply and therefore they are an attractive proposition for some customers. Interruptible supply contracts can be negotiated in a variety of forms, with a variable interruption period. Not surprisingly, the longer the permitted period of interruption the lower the unit price of the gas. With an interruptible supply tariff or contract, the customer is given short notice that the gas supply is going to be cut and must therefore be prepared to switch over to another fuel.

Table 2.6   Typical gas tariff for small customer

Standing charge per quarter	£ 9.57
Unit charge per kWh supplied	1.52 p/kWh

Table 2.7   Fuel oils and their properties

Class	Domestic heating oil C	Gas oil D	Light fuel oil E	Medium fuel oil F	Heavy fuel oil G
Viscosity (Redwood No.1 @ 100 secs.)	–	35	220	950	3500
Density at 16 °C (kg/l)	0.79	0.85	0.94	0.97	0.98
Pour Point (°C)	–	−6.7	17.0	21.0	21.0
Gross calorific value (MJ/l)	35.5	38.7	40.8	41.6	41.7

From Eastop and Croft (1990) *Energy Efficiency for Engineers and Technologists*, © Longman Group Ltd 1990, reprinted by permission of Pearson Education Ltd and Porges and Porges (1976) *Handbook of Heating, Ventilating and Air Conditioning* Newnes-Butterworths.

2.6 Fuel oil

Unlike gas and electricity, oil is not supplied under a tariff, but must be purchased on the open market. This means that the price of oil can be extremely volatile. Where possible it is worth using dual burners on boilers, so that if the price of oil gets too high, natural gas can be used as an alternative.

There are a variety of fuel oils on the market, ranging from free flowing domestic heating oil to thick heavy grade oils. Classification of fuel oils is usually made according to their viscosity (as shown in Table 2.7).

Unlike gas and electricity, which are relatively easy to handle, the use of fuel oil imposes restrictions. It is therefore important to consider the storage and handling facilities required, since these can affect purchasing arrangements and also impact on energy consumption. The storage volume in relation to average and peak consumption rates should be considered. Heavier grades of oil need to be heated in order to facilitate pouring, it is therefore worth considering reducing the volume that is kept heated, in order to reduce standing losses. Whilst it is normal to provide enough storage for approximately 3 weeks operation [9], in facilities which have an interruptible gas supply it may be necessary to install additional storage capacity.

Storage tanks in which heavier fuel oils are kept heated should be well insulated in order to minimize standing losses. In addition, the lower price of heavier grade oils should be set against the additional costs associated with handling the oil and maintaining it at a pouring temperature. For example, for Class G fuel oil a pumping temperature of 55 °C is required.

References

1. BP energy statistics 2001, BP Website (www.bp.com/downloads/702/BPwebglobal.pdf) (March 2002).
2. Campbell, C. J., Laherrère, J. H., George, R. L., Anderson, R. N. and Fouda, S. A. (1998). Preventing the next oil crunch. *Scientific American*, **278**(3), 59–77.
3. Seven Year Statement (1994). The National Grid Company.
4. Wendland, R. D. (1987). Storage to become rule, not exception. *ASHRAE Journal*, May.
5. Electricity supply in the OECD (1992). *International Energy Agency/OECD*.

6. Eastop T. D. and Croft D. R. (1990). *Energy efficiency for engineers and technologists.* Longman Scientific & Technical.
7. Natural gas transportation: Organisation and regulation. (1994). International Energy Agency/OECD.
8. Porges, J. and Porges, F. (1976). *Handbook of heating, ventilating and air conditioning.* Newnes-Butterworths.
9. Energy audits and surveys (1991). CIBSE Applications Manual AM5.

Bibliography

Eastop, T. D. and Croft, D. R. (1990). *Energy efficiency for engineers and technologists.* (Chapter 9), Longman Scientific & Technical.

Economic use of electricity in buildings (1995). Fuel Efficiency Booklet 9. Department of the Environment.

Energy audits and surveys (1991). CIBSE Applications Manual AM5.

Yuill, D. (1985). Understanding electricity costs. *Energy Manager's Workbook,* **2** (Chapter 2), Energy Publications.

3

Competition in energy supply

In this chapter the concept of competition in the energy sector is examined for both electricity and gas supply industries and the experience of electricity deregulation in the UK and USA is discussed in detail. The potential role of demand-side management is also investigated and comparisons are drawn between experiences in the UK and USA.

3.1 Introduction

It has traditionally been the case that gas and electricity utility companies, irrespective of ownership (i.e. state or privately owned), are natural monopolies, which are regulated by legislative measures. These monopolies evolved partly because of the high infrastructure costs associated with the transmission and distribution of gas and electricity, and partly because it was easier to manage and regulate utility companies which generated/supplied, transmitted and distributed electricity or gas. Indeed, it is difficult to imagine anything other than a monopoly, given that most buildings have only one physical connection to a gas pipe and another to an electricity cable. However, while monopolistic utility companies are relatively easy to control and regulate, they prevent competition in the energy market. Consequently, it is not possible to buy and sell 'bulk' energy in the same way other commodities are traded.

In recent years many governments around the world have begun to investigate alternative solutions which introduce competition into their respective electricity and gas supply industries. This has become possible because of various technical and financial advances made in the late 1980s and 1990s. The United Kingdom has been at the forefront in pioneering utility deregulation, and has completely restructured its utility sector. During the 1990s the UK deregulated first its electricity supply industry and then its gas industry, in a long and complex process, which at time of writing, is still ongoing. Such has been the radical nature of these changes that in many ways the UK has become the 'pilot study' for the rest of the world. Following the UK's lead a number of countries, including the USA, have deregulated (at least in part) their electricity supply industries and are developing new energy trading markets. In addition to the UK electricity spot market (i.e. the electricity 'pool'), four other 'pools' have so far been created in Europe; the Amsterdam Power Exchange (covering The Netherlands, Belgium and Germany), the Spanish Pool, the Swiss Pool and the Nordpool in Scandinavia (covering Norway, Sweden and Finland) [1].

3.2 The concept of competition

Consider the case of an organization which uses oil to heat its buildings. Under normal circumstances the organization will have a choice of competing fuel suppliers from whom to purchase oil. The organization can negotiate a bilateral supply contract with any one of these suppliers. If one supplier becomes too expensive, then the organization can simply switch to purchase oil from another supplier. If the general demand for fuel oil is high, then the suppliers will be able to raise their prices. Conversely, if demand is low then the price of oil will also be low. Thus a competitive market in fuel oil exists which reflects the demand for oil at any moment in time. As with any other commodity, the oil price will vary because customers have the ability to switch between suppliers. In addition, there is no cross-subsidy of one group of customers by another group of customers. Each fuel supply contract is negotiated on an individual basis between the parties concerned.

Now consider the same organization purchasing electricity under a tariff from a utility company. Since the electricity is supplied through cables owned by the utility company, the customer has no choice of alternative supplier and so the organization is compelled to purchase electricity at a price fixed by the utility company. As a consequence:

- *No competition exists*: The customer is in a weak position since electricity prices are fixed by the utility company.
- *No market exists*: Under a tariff, electricity prices are fixed, with the result that prices do not accurately reflect the fluctuations in demand for electricity. Although many tariffs do have reduced 'off-peak' elements these are at best only a crude indicator of market demand.
- *The potential for cross-subsidy exists*: The utility company may decide to offer lower electricity prices to its large industrial customers, and recoup some of its lost income by increasing the prices of its smaller domestic and commercial tariff customers. This is termed 'cross-subsidy', and effectively means that one group of customers is subsidizing another group.

While this monopolistic scenario may suit the utility companies, it does not benefit the customer. The utility companies are in a strong position and the potential exists for artificially high electricity prices. Lack of competition ultimately leads to:

- Manufacturing industry paying a high price for energy, with the result that the unit cost of production increases and the industry becomes less competitive.
- Utility companies becoming over-manned and inefficient.

It is therefore easy to see why many governments are reviewing the monopolistic position of their respective utility companies with a view to introducing a competitive energy market.

3.3 Competition in the electricity supply industry

While it is easy to state that competition in the energy market is a desirable thing, in practice it is difficult to achieve a truly competitive market amongst utility companies. Utility networks, be they gas or electricity, lend themselves to monopolies and are not naturally suited to competition. This is because it is impractical and prohibitively expensive to construct two or more sets of competing transmission/distribution networks. Given this, the simplest and easiest way to organize affairs is to have a 'vertically integrated' structure in which a single utility company is responsible for provision of supply. Figure 3.1 shows the structure of a typical vertically integrated electricity supply industry.

In a vertically integrated electricity supply industry the various utility companies have monopolies over their 'franchise' regions. Within its franchise region a utility company will be

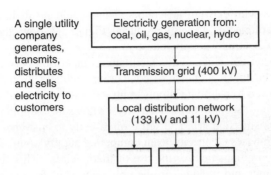

Figure 3.1   A vertically integrated electricity supply industry

responsible for generating, transmitting, distributing and supplying electricity to all its customers. Customers in the utility company's franchise region are forced to purchase their electricity from the regional utility company. Vertically integrated utility companies can exist in both the private and the public sectors. Their monopolistic position is derived solely from their physical location, which excludes competition and means that the utility company has a protected market. Under this scenario energy prices can easily become over inflated if the utility company is not tightly regulated.

In order to promote competition in the electricity supply industry it is necessary to create a market for the commodity which is flexible and yet, still robust enough to cope with wide fluctuations in demand. The market should:

- Allow various electricity supply companies and generators to compete with each other to sell electricity direct to customers.
- Allow customers to negotiate electricity supply contracts with various suppliers.
- Be transparent, so that generators, suppliers and customers can see that the market is fair and equitable.
- Create a 'spot market' which accurately reflects both demand for energy and cost of production. This spot market then becomes the market indicator of the real cost of production at any given point in time.
- Facilitate a future's market in electricity trading.

While the above points are relatively easy to achieve in a normal commodity market, they are not easily achieved in a market in which electricity is bought and sold. This is because electricity cannot be stored and must be generated only when it can be consumed. Any potential trading market in electricity must fully accommodate the physical constraints of an electricity supply system. As a result a truly competitive market in electricity is likely to be much more complex than a normal commodities market.

It is impossible to achieve a competitive market with a vertically integrated electricity supply industry. Instead a horizontally integrated structure is required. The introduction of a horizontally integrated electricity supply industry, in which the generating, transmission and distribution roles are all split up from each other, is the key to facilitating competition. By splitting up the roles it is possible to create competition between generators, who then have to bid in a 'spot market' for the right to supply electricity to the transmission grid. If the transmission company acts in a fair and independent manner, purchasing power at 'least cost', then any possible cartel should be eliminated. It then becomes possible for new 'independent power producers' to enter the market to compete with existing generators. This should result in a reduced cost to the customer for each unit of electrical energy produced. Figure 3.2 shows the structure of a typical horizontally integrated electricity supply industry.

Whilst the spot market described above facilitates competition between generators, it does not of itself offer the customer a choice of competing suppliers. In order to achieve this the customer must be allowed to negotiate supply contracts with individual energy suppliers. This is achieved by allowing 'second tier' electricity 'wholesale' suppliers to purchase 'bulk' electricity from the transmission grid and sell it on directly to customers. Under this arrangement the customer purchases electricity from competing supply companies, who pay a fee to the relevant distribution companies for the use of their 'wires'. This 'line rental' fee is then passed onto the customer and included in the unit price paid for the electricity. In order to ensure that true competition takes place, the 'line rental' fees should be transparent and equal for all potential electricity suppliers. These fees are usually fixed by some form of statutory regulatory mechanism.

This discussion indicates that facilitating competition in an electricity supply industry involves the setting-up of a complex structure, with many demarcation boundaries. Indeed, there is an inherent conflict of interests between the engineering and financial requirements of a horizontally integrated structure. The transmission company is primarily interested in procuring enough electrical energy from generators in order to meet the instantaneous demand on its grid. It seeks to procure this energy from the cheapest power producers and is not particularly interested in individual supply contracts. Conversely, customers, suppliers and generators are primarily interested in negotiating contracts which ensure secure supply and therefore are not interested in the transmission company's need to meet instantaneous demand. Satisfying these conflicting needs requires the setting-up of complex bidding, pricing and settlement mechanisms. It is the specific nature of these mechanisms and the efficiency with which they are applied which will ultimately determine the success or failure of any electricity market.

In addition to the complex financial and settlement mechanisms required to operate the market, suppliers need to know the 'real-time' electricity consumption of their contract (i.e. non-domestic) customers. This involves the installation of 'smart' meters which measure electricity usage every half hour, and can be read remotely and automatically. The data from these meters are transmitted to remote disseminated centres, from which relevant data are sent to all the parties involved in the supply contract. Contract customers may purchase or lease their metering equipment, but the installation and maintenance of these meters should be carried out by approved operators.

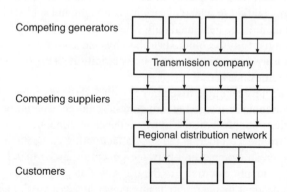

Figure 3.2   A horizontally integrated electricity supply industry

3.4 The UK electricity experience

Competition in electricity supply is still in its infancy and many protocols are not yet firmly in place. Most of the electricity power markets which exist around the world are only a few years old and even the UK market, established in 1990, is still undergoing major revisions. This makes it difficult to describe general rules which apply to all electricity markets. In the absence of any firm 'ground-rules' it is worthwhile looking in detail at the evolution of the competitive electricity market in England and Wales (the largest part of the UK), since this has been the 'template' for subsequent deregulation schemes in various parts of the world.

In 1990 in England and Wales a daily spot market known as the electricity 'pool' was created. The pool was administered by the National Grid Company (NGC), which owned and operated the transmission grid in England and Wales. Each morning the competing generating companies would submit 'bids' for their various generating sets to NGC for the following day's operation. Each bid included an offer price at which the generating company would be prepared to operate its various generating units for the following day. It also included a declaration of availability of generating plant for the following day. Once the generators had submitted their bids to the pool, the NGC examined its own demand forecast for the follow-ing day, and ranked each generating unit in order of price (lowest price first), so that finally a merit schedule was produced. This schedule was then published at approximately 15.00 hours, so that the generating companies were notified of the generating units required for the following day. As there was often considerable over-capacity in the system, any generating units for which the offer price was too high were either placed on standby, or excluded from the pool and forced to shut down.

As electricity cannot be stored it is essential that the controllers of the national transmission grid be able to bring on line (or down load) additional generating capacity at very short notice to cope with fluctuating demand. Figure 3.3 shows the national grid demand profile for a peak 'winter time' weekday, 29 November 1993 [2]. This graph shows that demand on that day varied considerably over the 24 hour period. To cope with increases in demand, generating units had to

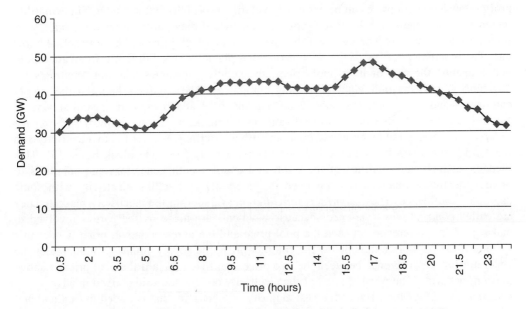

Figure 3.3   Demand experienced by National Grid, 29 November 1993[2]

be brought on line as and when they were required, but in strict accordance with their ranking in the daily pool merit schedule. In other words, generating sets which bid a low price were brought on line first, while the more expensive units had to wait until demand increases before they were allowed to generate. Consequently the pool price varied for each half hour period throughout the day. When demand was high, it generally followed that pool price would also be high. In this way the pool price reflected the demand on the transmission grid.

The bid price submitted for the most expensive generating unit brought on line to meet the demand in any given half hour period was known as the 'system marginal price' (SMP). For example, if the highest bid price accepted into the pool for the half hour period 11.00 to 11.30 hours was 2.5 p/kWh, then the SMP would be 2.5 p/kWh. It is important to note that it is the SMP, not the bid prices submitted by the individual generators, which became the basis for the eventual pool price for any given half hour, and that all the generators on line in that particular half hour were paid the 'pool purchase price' (PPP). Electricity supply companies and large consumers purchasing from the pool had to pay the 'pool selling price' (PSP). Not surprisingly PSP is always greater than PPP, the difference being an uplift to cover the pool operating costs. The electricity pool in England and Wales enabled a competitive market to exist amongst the generators, and gave the market as a whole an indication of the true costs of electricity production at any given time.

While an electricity pool facilitates competition between the various generators, it does not on its own provide the mechanism for promoting a competitive market amongst customers. In order to achieve this, 'second tier' electricity 'wholesale' supply companies must be allowed to purchase electricity from the transmission grid and sell it on directly to customers. These wholesaler suppliers negotiate bilateral contracts with the generating companies to purchase 'bulk' electricity at fixed rates, under a series of *contracts for differences* (defined later in this paragraph), and then sell it on to the customer at a marked-up price. These supply companies make their money by purchasing 'bulk' electricity from the generators at a low price and selling it on to their customers at a higher price. This involves considerable financial risk and the supply companies must negotiate contracts which ensure that they make a profit. However, pool price can be extremely volatile, especially in the winter. This volatility increases the element of risk for the supply companies if they purchase from the pool, with the result that they may lose money if they purchase at a high price and have to sell at a low one. The inherent volatility of the pool also makes planning ahead difficult. In an attempt to hedge against the risk of high pool prices, the supply companies take out 'contracts for differences' with the individual generating companies. The contracts between the supply companies and the generators operate outside the pool and operate in a similar way to 'futures contracts' traded in the world's commodity markets. Under a typical *contract for differences* a supply company would contract with a specific generator to buy electricity at a fixed price for a specific time period (usually on a daily five time block basis) [3]. This 'hedges' against the volatility of the pool, and enables both generators and suppliers to predict the future financial risk involved in generating and selling electricity with some degree of confidence. These *contracts for differences* underpin the electricity market. They are called *contracts for differences* because payments are made by the parties involved to make good the difference between the pool price and the agreed contract price. Under this system, if the pool price falls below the contract price, the supply company remunerates the generator for the difference between the two prices, and vice versa if the pool price is above the contract price. The price of most of the electricity bought and sold is fixed in advance by *contracts for differences*. Hence the vast majority of electricity that is traded in England and Wales is purchased outside of the electricity pool.

3.4.1 The evolution of the UK electricity market

The electricity pool described in Section 3.4 has become the basis for a number of other trading pools set up during the 1990s in Europe. However in the UK, during the late 1990s concerns were expressed that the pool system:

- Favoured the large generating companies; indeed there was suspicion that these companies were able in some way to control the pool price.
- Inhibited the introduction of new independent energy traders into the market.
- Inhibited the negotiation of bilateral electricity supply contracts between various parties.

The last point is an important one. In most trading deals the customer can state the price at which they wish to purchase a commodity and this has an influence on the overall market price. However, under the pool system the 'market price' (i.e. the pool price) was wholly determined by the sellers (i.e. the generating companies). The pool could therefore be viewed as being in some way only 'half a market' [4].

As a result of the concerns stated above, the UK completely restructured its electricity trading arrangements in 2000 and introduced the 'New Electricity Trading Arrangement' (NETA) [5]. This new arrangement abolished the old centrally regulated pool in favour of a 'free-market' approach which allowed a series of 'power exchanges' (i.e. electricity commodity markets) to be established; the hope being that the exchanges and brokers would create forwards, futures, and short-term bilateral markets. The intention was that the true price of electricity would become established through the power exchanges in much the same way that the commodity markets fix the price of other traded commodities. However, while the power exchanges can facilitate trade in 'bulk' energy, there is no way in which they can satisfy the physical engineering requirements of NGC (the operators of the transmission grid), who need to predict accurately at any point in time the demand on their network. Because electricity cannot be stored, the NGC must bring on-line more generating capacity as demand rises, otherwise power cuts will occur. Clearly, no commodity market can solve this problem alone! So the NETA arrangements were designed to operate in parallel with the new power exchanges, so that every time a bilateral contract is signed between a generator and a supplier they are required to inform NGC (or its settlements agent) of the quantity of electricity traded and the duration of the contract. It should be noted that parties are not required to notify NGC of the price paid for the electricity. All the supply companies and the generating companies are also required to notify NGC in advance of their expected operating levels for the following day. So by 11.00 hours on the day before trading, both the generators and the suppliers must submit to NGC their forecasts of demand, on a minute-by-minute basis, for the day ahead. They can do this because all parties know the quantity of electricity they have contracted to supply or purchase for the following day. By 'gate closure' (i.e. 3.5 hours before real time) the suppliers and generators must submit finalized demand forecasts to NGC. In this way NGC can effectively manage the transmission grid and inform the individual generators of the generating plant that will be required for the following day.

In theory the demand profiles predicted by the supply companies should exactly match the generation profiles predicted by the generating companies. In reality this never happens because it is difficult to accurately predict demand for electricity on a daily basis. A large number of factors influence electricity consumption, including weather and television scheduling. Since many variables influence electricity consumption, it is inevitable that the true demand for electricity will vary from the demand predicted by the supply companies. This means that NGC will have to bring on-line (or take off-line) at short notice, additional generating plant in order to cope with variations from the predicted values. This of course incurs additional expense on behalf of the generating companies who have either to bring on-line extra plant or lay-off generating plant which it had planned to operate. These 'imbalance costs' (i.e. costs incurred due to deviations from bilateral supply contracts) are calculated by NGC through a complex series of counter 'bids' and 'offers' made by both the generators and the suppliers. In this way NETA determines only the unit price of electricity which is 'traded' at the margins (i.e. outside of the power exchanges). It is intended that the 'imbalance' electricity costs will be higher than the 'bulk' electricity price, thus encouraging both the generators and the suppliers accurately to forecast predicted demand.

From the discussion above it is evident that in order to accommodate the engineering constraints of a transmission grid and facilitate a commodity market in electricity, extremely complex trading arrangements must be set up. Given this, it is understandable that competition in the electricity supply sector has been slow to evolve. Indeed, it and would have been impossible without recent rapid advances in information technology (IT) in general and the Internet in particular. Without these IT advances, it would be impossible to rapidly transfer the large amounts of data associated with the bidding process to the many parties involved in a power exchange.

3.4.2 The Californian experience

From the discussion in Section 3.4.1, it is clear that facilitating a true competitive market in electricity is an extremely complex process. Indeed, the electricity supply industry is of such strategic importance that if the deregulation process goes wrong, it can have a catastrophic effect on the whole economy. With this in mind, the experience of the Californian electricity supply industry should be a salutary lesson to all legislators who might be considering deregulating their utility sector. In January 2001 large parts of the state of California suffered major power-cuts, not because of any technical failures, but as a direct result of poorly thought out legislation [6].

In 1996, the California Assembly voted to deregulate the state's electricity supply industry and to dismantle what was considered to be a government regulated monopoly [7]. Prior to deregulation, the state had a vertically integrated electricity supply industry, with a number of investor-owned utility companies owning and operating their own power stations, transmission grids and distribution networks. With deregulation, a non-profit making organization, the California Power Exchange was established and the following changes were made:

- Operational control of the transmission grids was transferred to a single Independent System Operator who became responsible for the management of the system.
- The investor-owned utility companies, such as Southern California Edison and Pacific Gas and Electric, were forced to sell most of their power stations to other unregulated private companies. This forced the major utility companies to purchase wholesale electricity through the California Power Exchange.
- The California Power Exchange acted as a wholesale commodities market, through which all the state's electricity was bought and sold. An auction process therefore set the price of wholesale electric power.

The investor-owned utility companies did however, retain ownership and control of their distribution networks.

By making these changes the California legislature created a classic model for a competitive, deregulated electricity supply industry. However, there were two critical factors which were to have a significant influence on the events that were to follow:

1   While deregulation forced the utility companies to purchase their power on the open market and pay market prices, it prevented them from passing on any increases in the cost of whole-sale electricity to their customers until at least 31 March 2002 [7].

2   Because of environmental concerns the state authorities prevented the building of new power
    stations. For 20 years or more, there had been no significant increase in California's
    generating capacity, despite the fact that demand for electricity in the state had been growing
    at approximately 2% each year [7].

These two critical factors were to have disastrous consequences for California in general and its
electricity supply industry in particular. What the state legislature had done was to force the utility
companies to buy wholesale electricity on the open market, which can be extremely volatile, while
at the same time effectively fixing the price at which the utilities could sell electricity to their
customers. The failings of this strategy were compounded by the fact that there was little excess
generating capacity in the system. Without excess capacity there was little competitive pressure to
keep wholesale prices low. As a result during the summer of 2000, when demand for power peaked,
the utility companies urgently needed power from the electricity wholesalers and generating
companies, who promptly raised their prices. Bulk electricity prices rose steeply, with the average
price of electricity bought through the Power Exchange rising from approximately $30 per MWh in
January 2000 to $330 in January 2001 [8]. In fact, in December 2000 the price reached a peak of
$1400 per MWh [7]. Unable to recoup these inflated costs from their customers, the utility
companies, not surprisingly, started to lose money. They rapidly ran out of money, with the two largest
utilities, Southern California Edison and Pacific Gas and Electric, claiming that by January 2001 their
combined losses exceeded $9 billion [7]. Indeed, Pacific Gas and Electric filed for bankruptcy in April
2001 [9]. The financial difficulties of the utility companies had two direct consequences:

1   The banks became very reluctant to lend more money to the cash-starved utility companies,
    who were rapidly becoming insolvent.
2   The wholesale and generating companies became reluctant to sell electricity to utility com-
    panies which were obviously in financial difficulties.

Faced with such high financial losses and not wanting to lose any more money, the utility companies took the only course of action available to them: they stopped purchasing electricity and
the state of California suffered major power-cuts. The state authorities then had to step in and try
to pick up the pieces and sort the mess out.

    The sorry state of affairs that occurred in California graphically highlights the major problems
which can occur if all the issues involved in deregulation are not thought out in advance. Clearly,
the combination of a shortage in generating capacity and an unregulated wholesale market,
facilitating what is in effect an energy cartel, is a recipe for disaster.

3.5 Competition in the gas market

In many ways facilitating competition in the gas market is similar to the electricity market. As
with electricity supply, horizontal integration is the key to a competitive gas market. However,
there are a number of fundamental differences which make trading in natural gas much simpler
than trading electricity:

* Natural gas is not generated; it is pumped out of oil and gas fields at sea or on land and sold
  to licensed shippers (i.e. wholesale supply companies) who sell it on to customers.
* Unlike electricity, natural gas can be stored to a limited extent.
* Demand for natural gas is very seasonal.

Given the differences between the nature of gas and electricity, a relatively simple horizontally
integrated model is required to facilitate a competitive market in gas (as shown in Figure 3.4).

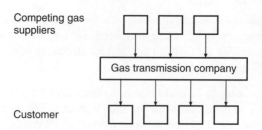

Figure 3.4 Horizontally integrated gas supply industry

Because there are only three parties involved in the process and also because gas can be stored, the whole structure is much simpler to control and operate than that of an electricity supply industry. However, in order to ensure that the system functions in a fair and equitable manner it is important that the gas transmission company charges equal transportation fees to all suppliers and that all fees should be transparent.

Under a horizontally integrated gas supply structure, individual customers are free to negotiate bilateral supply contracts with various competing suppliers. The price paid by the customer is the price the supplier pays for the gas at the 'beach head' plus the cost of transportation plus the supplier's profit. However, the price paid by the customer is mainly affected by the cost of gas at the beach head.

It is the responsibility of the gas transmission company to balance supply and demand on its network on a continual basis. If too much gas enters the network then it must be stored in underground caverns or gasometers. Conversely, if too little enters the network, then gas from the storage vessels will have to be utilized. Suppliers therefore have to ensure that the gas they put into the network is roughly equal to the gas that their customers use. If they miscalculate either way by too great a margin, then the transmission company will levy a penalty charge on them.

3.6 Load management of electricity

From the discussions in Section 3.4 it can be seen that the 'true' cost of electricity production varies with demand on the network, and that through the use of pricing mechanisms such as the 'pool' it is possible to introduce real-time electricity pricing. Under this scenario *when electricity is consumed* becomes as important as *how much electricity is consumed*. Those customers who have the ability to manage their electrical load should thus be in a good position to reduce energy costs.

An ability to manage electrical load not only reduces customers' electrical costs, it also enables them to negotiate more competitive electricity supply contracts. If a potential customer wishes to negotiate a supply contract, they will need to furnish potential suppliers with the following information:

- The annual consumption of electricity in kWh.
- The maximum demand in kW.
- The load factor.

The load factor for any given period represents the percentage of time for which plant and equipment operates during that period. It can be calculated as follows:

$$\text{Load factor} = \frac{\text{Energy consumed (kWh)}}{\text{Max. demand (kW)} \times \text{Time period (h)}} \times 100$$

Table 3.1   Typical load factors for a variety of applications [10]

Type of organization	Load factor
24 hour operation	0.7–0.85
2 shift system	0.45–0.6
Single shift system	0.25–0.4
Modern hotel complex	0.5–0.6
Hospital	0.6–0.75
Retailing	0.3–0.4
Catering business	0.3–0.5

Table 3.1 shows some typical load factors which might be expected for a variety of types of organization [10]. Buildings such as air conditioned commercial offices, with a high daytime peak and a low night time demand, will exhibit a low (i.e. poor) load factor. At the other extreme factories which operate a 24 hour shift system will exhibit a high (i.e. good) load factor.

From the utility companies' point of view organizations which possess a high load factor are potentially more desirable customers, since they will be buying more electrical energy for a given amount of investment in generation and distribution equipment. Customers who possess high load factors should therefore expect to negotiate better supply contracts than those with low load factors. This provides great potential benefit to contract customers who possess the ability to load shift from day to night by using technologies such as ice thermal storage (see Chapter 11). This should be particularly true for office buildings which would otherwise exhibit a very poor load factor.

3.7 Supply side and demand side

The collective term for the operations performed by utility companies is the 'supply side', whereas energy consumption by customers is referred to as the 'demand side'; so named because customers create a demand for energy which is then supplied by utility companies. These concepts are illustrated in Figure 3.5.

Consider the case of an electricity utility company which experiences an overload of its system during the daytime in the winter months. The company cannot meet the increase in demand with its existing generating plant and is therefore faced with the choice of either, building more power stations or encouraging its customers to consume less electricity and thus reduce electrical

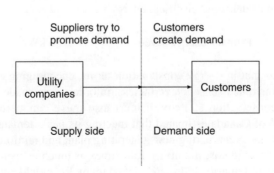

Figure 3.5   Concept of supply side and demand side

demand during the daytime. The former solution is a 'supply-side measure' since the solution lies wholly with the utility company (i.e. on the supply side) and the latter is termed a 'demand-side measure' since the solution to the problem lies with the customer. The demand-side solution could be achieved by introducing an electricity tariff offering lower unit charges to customers who are prepared to switch their electricity consumption from the daytime to the nighttime. Through management of the 'demand side' in this way it is possible for utility companies to utilize their resources efficiently and thus achieve substantial cost savings. Demand-side measures are therefore concerned with direct intervention in the customer's end use of electricity by the utility company, in a way which affects the planning of the utility company's infrastructure.

Traditionally electricity utility companies have tended to rely on supply-side measures to shape their businesses; that is, the utility companies have tried to influence the way in which their customers use electricity from the supply side of the meter, and have provided the infrastructure to meet the predicted demand. However, in recent years, both in the UK and the USA, there has been increasing interest in the use of demand-side measures.

3.8 Demand-side management

The concept of demand-side management (DSM) (sometimes referred to as 'least cost planning') was pioneered in the USA during the 1980s, where it has since become an influential force. In some parts of the USA the electrical demand can increase by as much as 40% during the summer months, due to the use of air conditioning equipment [11]. There is also stiff legislative opposition from the Public Utility Commissioners to the construction of new power stations. Faced with this situation many USA utility companies have introduced DSM programmes to encourage customers to conserve energy, and persuade as many as possible to shift their daytime load to the nighttime. In the USA DSM programmes include such measures as financial support for feasibility studies, free advice on techniques, capital grants towards the cost of new equipment, and even the free issue to customers of low energy light bulbs. Many utility companies in the USA have found it more economical to persuade their customers to conserve energy, rather than be forced to build new generating plant. A typical example of this is that of Pacific Gas and Electricity, which in 1985 announced that it intended to 'build' a new power plant; a 1000 MW conservation power plant. In other words they intended to buy extra efficiency improvements which would reduce their peak demand by 1000 MW [12].

Simple analysis of energy consumption demonstrates the great benefit of encouraging energy conservation over the construction of new generating plant. If it assumed that a typical thermal power station has an efficiency of 35%, then the overall primary energy saved through the conservation of 1 kWh of delivered electrical energy is:

$$\text{Primary energy saved} = \frac{1}{0.35} = 2.86 \text{ kWh}$$

From this it is obvious that in energy conservation terms, encouraging customers to conserve electrical energy makes much sense. Nevertheless, in order to persuade the utility companies to adopt an energy conservation strategy, it must also make commercial sense. In the late 1980s, Ontario Hydro of Canada estimated that meeting its peak demand obligations through supply-side measures (i.e. constructing new generating plant and reinforcing transmission and distribution networks) would cost the utility four times as much as using demand-side measures. [13] The findings of Ontario Hydro are backed up by Rosenfeld and de la Moriniere [14] who demonstrated in 1985 the cost of constructing new generating capacity to be in the region of $1200–$1500/kW, which compared very poorly with the maximum of $400/kW of electricity saved which could be achieved by using an ice storage system. It is therefore clearly in the interests of vertically integrated utility companies, such as those that exist in many parts of the USA and Europe to encourage the installation of DSM technologies. To this end, many of the utility companies in the USA offer substantial capital incentives to building users to install technologies such as low energy light fittings and ice thermal storage [12].

Although DSM has become an influential force in the USA, its country of origin, the UK has been slow to adopt it. The UK does not suffer from a shortage of generating capacity, as is the case in some parts of the USA. It also experiences a winter peak, unlike many states in the USA. In addition, in England and Wales the electricity supply industry is not vertically integrated as much of the USA still is, thus making comparisons between the two countries very difficult. However, despite the obvious differences between the electricity supply industries in the USA and England and Wales, the regional distribution companies in the UK have recently become interested in DSM, since it is one method by which they can significantly reduce the demand on their cables and transformers, and thus reduce their operating and capital investment costs.

Because of the complex nature of the UK's horizontally integrated electricity supply industry, the role of DSM in the UK is somewhat ambiguous. In theory the widespread introduction of DSM should:

- Produce a reduction in the fuel burnt at power stations.
- Cause the deferral of the capital and financing costs of new power station construction.
- Cause a reduction in distribution losses.
- Result in the possible deferral of distribution reinforcement.
- Cause a reduction in transmission losses.
- Result in the possible deferral of transmission reinforcement associated with both new power plants and increased loads.
- Lead to a reduction in the emissions of CO_2, SO_2 and NO_2 from power stations.

While at first sight all the above points seem to indicate that there is a strong case for implementing DSM policies in the UK, further analysis casts doubt on the validity of the statement above. In theory all parties in the UK electricity supply industry benefit from the introduction of DSM. Yet, because of the fragmentation of the industry due to horizontal integration it is difficult to initiate and coordinate an effective DSM policy. For example, who will pay for a DSM policy? Are the regional distribution companies going to pay for a policy which arguably gives greatest benefits to the generators and the NGC? It is also difficult for the competing generators to initiate DSM, since they have no 'captive' market and they have little direct influence over the end users. Also, the structure of the electricity market is such that individual generators are always seeking to generate as much electricity as possible. The benefit to the generators through the implication of a DSM policy is dubious to say the least, since there is over capacity in the system, and every generator benefits from higher electricity prices when demand is high. Therefore, for DSM to succeed in the UK it must benefit both the regional distribution companies and their customers.

3.8.1 The USA experience

The US-based energy research body the Electrical Power Research Institute (EPRI) defines DSM as:

The planning, implementation and monitoring of utility activities designed to influence customer use of electricity in ways that will produce desired changes in load shape [15]

In the USA, DSM programmes are often initiated by the Public Utility Commissioners who are intent on minimizing the construction of new generating plant. Utility companies are required to demonstrate to the Commissioners that their proposed course of action is the least expensive option for supplying customers with electricity. The onus is therefore on the utility companies to reduce demand rather than build more power stations. In some states in the USA, utilities are even being awarded bonuses for implementing DSM programmes.

Although DSM programmes in the USA have been initiated as a result of social concern and regulatory pressure, it is the potential for profit to the utility companies that has driven such programmes. In the USA the utilities are permitted to over-recover the costs of DSM programmes through increases in electricity prices. Consequently, the utilities receive a greater marginal return from demand-side measures than they would from supply measures. This has resulted in DSM programmes in North America being used on a large scale. Many North American utility companies spend more than 5% of their total turnover on investment in DSM. Table 3.2 shows the investment levels and targeted energy savings for some DSM programmes, operated by a variety of North American utility companies [16].

Although some of the DSM programmes included in Table 3.2 have not proved to be cost effective, many of the utility companies have reported that their DSM programmes have proved less expensive in total cost terms, when compared with the costs avoided on the supply side. These findings even applied in circumstances where the utility company had an excess of generating capacity.

When a DSM policy is introduced a utility company avoids generating costs, network losses, some administration charges, and may avoid capital expenditure on network reinforcement and expanding generating capacity. However, it also sells less electricity and is therefore liable to a loss of revenue through implementing a DSM programme. To avoid this situation some form of 'balancing' mechanism must be provided to ensure that the utility company does not lose revenue. In the USA, this balancing mechanism is provided by a regulator, who approves an

Table 3.2   Examples of North American utilities' expenditure on DSM [16]

Utility company	Current expenditure ($ Millions)	Target GWh savings	Target MW Savings	MW savings as % of projected peak	Target year for savings
BC Hydro	66	4491	1266	9.4	2000
Hydro Quebec	251	9289	5065	13.2	2000
Manitoba Hydro	8	931	255	4.7	2000
Ontario Hydro	377	14911	5200	16.0	2000
Consolid. Edison	76	7120	2500	22.5	2008
Florida P & L	66	2800	1884	8.7	1999
Long Island	33	2840	589	11.4	2008
Nevada Power	5	190	147	5.2	2007
New York State	25	2790	846	18.9	2004
Niagara Mohawk	37	2680	849	12	2008
Orange & Rock.	8	191	122	7.6	2008
Pacific G & E	120	5760	2270	11.1	2001
Rochester G & E	7	876	186	10.7	2009
Southern Calif.	107	5170	2780	11.2	2009
Wisconsin Elec.	40	1260	290	5.6	2000

increase in tariffs for all customers, subject to the utility company demonstrating that the 'average' customer receives an overall reduction in energy costs [17].

In recent years the electricity supply industry in the USA has undergone major restructuring in order to facilitate wholesale trading in electricity in a similar way to the industry in the UK [18]. Despite the uncertainty that surrounds this change, the industry in the USA reported that in 1999, a total of 848 electricity utilities had DSM programmes and of this number the 459 'large' DSM programmes resulted in a 50.6 billion kWh energy saving [19].

3.8.2 The UK experience

Unlike North America, where DSM programmes have become commonplace, DSM in the UK is still in its infancy. Under the old state-owned electricity supply industry, one of the few examples of a DSM policy in the UK was the introduction of the Economy 7 tariffs which were used in conjunction with night storage heaters. Over many years under the nationalized regime, night storage heaters were heavily marketed, the main objective being:

- To achieve better utilization of the nation's generating plant.
- To utilize the electricity distribution network more efficiently.
- To raise useful revenue for the regional electricity boards by selling the night storage heaters to the public.

The marketing of night storage heaters was an extremely successful policy – perhaps too successful. Analysis of the pool price profile for an average weekday in December 1992 (see Figure 3.6) shows that the pool selling price for some of the nighttime is actually greater than the daytime (office hours) price. This was because of the generating capacity required at nighttime to satisfy night storage heaters. However, this high nighttime pool selling price was not reflected in the price paid by tariff customers, typically between a third to a half of the day time price, for both domestic 'Economy 7' customers and a commercial maximum demand tariff customers. In the case of 'Economy 7', the off-peak price was set to compete with gas central heating in the domestic market. As a result the users of night storage heaters were in fact being subsidized by other customers who have to pay higher daytime prices.

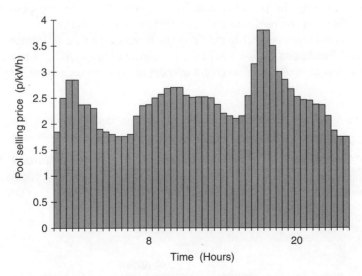

Figure 3.6   Average weekday Pool Selling Price December 1992

The intensive marketing of night storage heaters meant that in some areas of the UK, the regional distribution networks experienced high nighttime peaks. This caused problems and resulted in a number of regional distribution companies (who were also electricity suppliers) marketing flexible off-peak domestic tariffs. These flexible off-peak tariffs were designed to replace the old monolithic 'Economy 7' tariff, and offered customers 10 hours of off-peak electricity compared with the old 7 hour period [20]. A sample of one of these flexible tariffs is shown in Table 3.3 from which can be seen that the utility company is trying to utilize more effectively the troughs in the UK's daily demand profile, which generally correspond to periods when electricity prices are low.

To the regional distribution companies these flexible tariffs have a number of advantages. They shift much of the off-peak period from its 'traditional' nighttime slot, to the daytime and evening periods, so that troughs in the daytime demand can be exploited. They also have inherent flexibility which allows the utility company to control the precise start and stop times of the 'off-peak' periods, and allows these to be varied from day-to-day.

The regional distribution companies receive two major benefits from these flexible tariffs:

1 They achieve better utilization of their distribution networks, and avoid capital expenditure on network reinforcement.
2 If they are also a supply company, the regional distribution company can purchase electricity from the generators at periods when prices are low, and sell it on to their customers for heating purposes at the standard tariff price. Consequently, they have more scope for increasing profit margins in their supply business.

To implement flexible tariffs such as the one outlined above involves the installation of complex metering equipment, which is capable of both recording the electricity consumption at the various periods of the day, and also of receiving switching signals from the utility company concerned, to activate the 'off-peak' period on the meter. To achieve this in the domestic market the utility companies offering these tariffs have to use a radio tele-switching system.

If the subject of night storage heaters is set aside, DSM in the UK is being driven primarily by those regional distribution companies which are experiencing network problems [21, 22]. The position of the generators towards DSM is ambivalent, since it is unclear how they benefit commercially. Therefore, the potential benefits of DSM in the UK are perceived to lie in enabling the distribution companies to optimize their existing networks.

Electricity companies always seek to maximize their returns on their investment in generation, transmission and distribution equipment. In the past, increasing electricity demand has ensured that whenever a system needed reinforcement in order to maintain security of supply, the capital

Table 3.3   East Midlands electricity 'heatwise' tariff 1 May 1992 [20]

Off-peak supply is available for 10 hours	Monday to Friday	Saturday and Sunday
Five hours continuously during night	00.00–7.00	00.00–8.00
Three hours continuously during afternoon	13.00–16.30	13.00–17.30
Two hours continuously during evening	17.30–22.00	17.30–22.00
Standing quarterly charge	£3.90	
Unit charges:		
Off-peak	2.90 p/kWh	
Peak	7.64 p/kWh	

investment could be recouped from increased electricity sales. Before deregulation, the electricity supply industry used vigorously to promote the use of electricity in the hope of maximizing sales. This situation has however changed. The electricity market in the UK is a mature one. Sales of electricity have steadied and predicted growth is low. In some areas electricity sales are static or even declining. Distribution companies cannot look to increased sales to finance system reinforcement. Under this scenario DSM becomes an important option which the distribution companies must consider.

From the position of the competing generators in the UK, it is unlikely that DSM is going to gain much support. Electricity prices tend to be high when demand is high. Therefore all the generators benefit from high demand. From a generator's point of view, DSM can be viewed as a competitor since it reduces electricity sales.

DSM programmes cost money to implement, especially if they involve capital grants to customers to purchase energy efficient or load shifting equipment. Therefore the utility companies need some mechanism to recoup investment costs. In the USA, utility companies are allowed to increase tariff prices to all their customers to pay for DSM programmes. In effect the ordinary customers of the utility companies are subsidizing those customers benefiting from the DSM measure. In the UK the regulatory authorities will not allow this approach to paying for DSM, since it both distorts the market and is 'unfair' on franchise customers. Indeed, the regulatory authorities in the UK appear to be opposed to the widespread adoption of such schemes on the grounds that they represent a cross-subsidy between customers. The distribution companies must therefore recoup their DSM programme costs from those customers who benefit directly from it.

References

1. European electricity deregulation progress. www.commodities-now.com.
2. National Grid Company (1994). Seven Year Statement.
3. Business focus (1992). National Power Plc. Issue 3, July.
4. Littlechild, S. (1993). Demand side bidding in the Pool. OFFER, 5 May.
5. Haigh, R. (2000). An overview of the New Electricity Trading Arrangements V1.0. OFGEM and DTI, 31 May.
6. Campbell, D. (2001). Blackouts bring gloom to California. *The Guardian*, 19 January.
7. Feldman, C. (2001). The California power quagmire CNN.com, January 4. http://www.cnn.com/SPECIALS/views/y/2001/01/feldman.power.jan3. 29 March.
8. Feldman, C. (2001). California power crisis sends shock waves nationwide. CNN.com In-Depth Specials – Power Crisis. http://www.cnn.com/SPECIALS/2001/power.crisis. 29 March.
9. National Energy Information Center Subsequent Events – California's Energy Crisis (2001). http://www.eia.doe.gov/cneaf/ electricity/califo rnia/subsequentevents.html. 12 July.
10. Forrester, R. (1993). Hard Bargaining. *Electrical Review*, 9–29 July, pp. 32–4
11. Wendland, R. D. (1987). Storage to become rule, not exception. *ASHRAE Journal*, May.
12. Oliver, D. (1986). The Energy Efficient Way of Life. *CIBSE Journal*, June, p. 40.
13. Redford, S. J. (1995) Demand Side Management – Intervention across the meter. *CIBSE Journal*, June, p. 25.
14. Rosenfeld, A. and de la Moriniere (1988). The high cost-effectiveness of cold storage in new commercial buildings. *ASHRAE Transactions*, p. 818.
15. Duguid, T. and Mee, C. A. (1994). The drivers behind the UK experience of DSM. *Power Engineering Journal*, October, pp. 225–28.

16. Demand Side Measures – A report to the Office of Electricity Regulation (1992). LE Energy Ltd. SRC International ApS, 12 October.
17. Redford, S. J. (1994). The rational for demand-side management. *Power Engineering Journal*, October, pp. 211–17.
18. U.S. Electricity utility demand-side management: Trends and analysis. DOE (USA).
19. Electricity utility demand-side management (1999). DOE (USA).
20. Heatwise (1992). East Midlands Electricity Plc. 1 May.
21. Savings on demand. (1995). *CIBSE Journal*, June, p. 29.
22. Stanway, J. (1994). The Holyhead Experience. *Power Engineering Journal*, October, p. 221.

4

Energy analysis techniques

Energy data may be collected for buildings either through audits or as a result of continuous monitoring programmes. However, the data on their own are relatively useless unless they are interpreted in a correct manner. Analysis must therefore be carried out on any raw data that are collected in order to identify trends and areas in which improvements can be made. There are various analysis techniques which can be employed in order to interpret energy data. This chapter introduces the data analysis techniques which are commonly used in energy management.

4.1 Introduction

There are a wide variety of statistical and numerical analysis techniques which can be employed in order to understand why energy is being consumed in a particular fashion. Some of the analysis techniques are fairly simple and can be done using simple hand calculations, while others are more complex and may require the use of computer software. The purpose of this chapter is to introduce the various analysis techniques and explain the practice and theory involved in each.

4.2 Annual energy consumption

Probably the simplest analysis that can be undertaken is to produce a percentage breakdown of annual energy consumption and cost data. This is a useful technique which enables the overall energy performance of a building quickly and easily to be assessed. The analysis of annual energy consumption should be performed as follows:

(i) Convert all the energy consumption data into standard units (usually the kWh) using the standard conversion factors shown in Table 4.1 and the gross calorific values shown in Table 4.2.

(ii) Produce percentage breakdowns of the total consumption and cost of each energy type, and determine the average unit cost per kWh for each.

(iii) Compile a table similar to the example shown in Table 4.3 showing the total annual energy consumption, cost and percentage breakdown of each fuel type.

Table 4.1   Energy conversion factors

From	Multiply by factor	To
Therms	29.306	kWh
MJ	0.2778	kWh
GJ	277.778	kWh

Table 4.2   Typical gross calorific value of fuels [1]

Fuel type	Typical gross calorific value
Electricity	1 kWh
Natural gas	1.01 therms/100 ft^3
Gas oil (Class D)	38 MJ/litre
Heavy fuel oil (Class G)	42 MJ/litre
Coal	27–30 GJ/tonne
Propane	92.6 GJ/m^3
Butane	49.3 GJ/m^3

Table 4.3   Table of annual energy input for 1998/99

Energy type	Purchased units	Consumption (kWh)	(%)	Cost (£)	(%)	(p/kWh)
Electricity	61 500 kWh	61 500.0	26.0	3075.00	52.58	5.00
Gas	146 800 kWh	146 800.0	62.0	2231.36	38.16	1.52
Oil-class D	2700 litres	28 500.0	12.0	541.52	9.26	1.90
Totals	–	236 800.0	100.0	5847.88	100.00	2.47 (av.)

(iv) Produce pie charts similar to those shown in Figure 4.1 to show graphically the energy and cost contributions of each energy type.

(v) Where historical energy data are available, comparisons should be made in order to identify any trends, as illustrated in Table 4.4.

It is important to note that although the simple analysis described above may produce energy cost breakdowns and identify possible trends, no allowances have been made for variable factors such as the weather, which may influence the energy performance of the facility. It should therefore be viewed as a rather crude analysis technique, and should not be used when comparing the energy performance of one building against another. If comparison between buildings is required then a more sophisticated analysis approach such as that described in Section 4.3 is required.

4.3 Normalized performance indicators

A good indication of the energy performance of a particular building can often be gained simply by comparing actual annual energy consumption and costs with those achieved by buildings of a similar type and function. However, there are a number of inherent problems which must be overcome when comparing the energy performance of one building with another of a similar type in a different location.

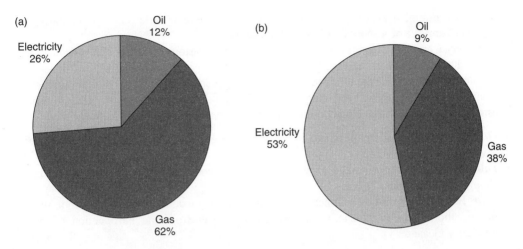

Figure 4.1 Energy consumption (a) and costs (b)

Table 4.4   Changes in annual energy use

Year	Consumption (kWh)	Change (%)
1994/95 (base)	201 456.4	n.a.
1995/96	197 562.2	−1.9
1996/97	203 216.2	+0.9
1997/98	220 403.5	+9.4
1998/99	236 800.0	+17.5

- The buildings may be of different sizes.
- The two locations may have different climates. This will influence the amount of energy consumed by the two buildings.
- The two buildings may experience different levels of exposure, which will influence building heat loss.
- The two buildings may experience different operating hours.

In order to overcome these inherent problems it is necessary to correct the building energy consumption data to allow for variables such as weather and occupancy patterns. The concept of the 'normalized performance indicator' (NPI) was developed to address these problems. NPIs enable the energy performance of particular buildings to be compared with others of a similar type and function. They also enable the overall energy performance of a building to be compared with standard energy 'yardsticks' for various building types (see Table 4.6).

Many countries around the world have national energy yardsticks for various building types. These yardsticks are determined by statistical analysis of the results of surveys of measured energy consumption. Energy yardsticks are usually quoted in kWh/m² of floor area per annum, although some prefer to use kWh/m³ of building volume per annum. Yardsticks provide a useful guide against which buildings can be measured. It should be noted that yardsticks are designed to provide useful guidance when establishing priorities; they should not be taken as absolute values to be achieved. In order to determine the actual potential for improving energy performance in a particular facility, it is necessary to undertake further detailed energy surveys. It is also important to remember that performance indicators are only intended to allow comparisons to be made between similar types of building. For example, there is little point in comparing the energy performance of a school with that of a hospital.

The precise energy yardsticks which should be used in various countries will obviously vary widely with the climate experienced and the nature of the construction used. Nevertheless, the principles, which underlie the production of NPIs, are generic and can be applied worldwide. For ease of reference, in this text the energy yardsticks and data applicable to the UK are used to illustrate the process. In the UK, building energy performance is classified as follows:

- *Good*: Generally good controls and energy management procedures although further energy savings are often possible.
- *Fair*: Reasonable controls and energy management procedures, but significant energy savings should be achievable.
- *Poor*: Energy consumption is unnecessarily high and urgent action should be taken to remedy the situation. Substantial energy savings should result from the introduction of energy efficiency measures [1].

Buildings which exhibit a 'poor' performance are most likely to offer the best energy management opportunities, but energy improvements should also be possible for those buildings classified as 'good'.

These broad categories can be assigned to most buildings for energy performance purposes. The process whereby NPI can be determined and building energy performance categorized is based on the CIBSE method [1] and is as follows:

(i) Establish the total building energy use in standard units as described in Section 4.2.
(ii) Ascertain the annual energy use for space heating. This can be determined either by separate sub-metering (see Chapter 5), or by using the analytical techniques described in Section 4.5. If this is not possible, the percentage breakdown data shown in Table 4.5 can be used as an approximate estimation.

Table 4.5   Proportion of fuel used for space heating and hot water production which is assumed to be attributable to space heating [1]

Building type	Proportion of fuel used for space heating and hot water attributable to space heating (%)
School	75
Hospital, nursing home	50
Other health care	75
Further/higher education	75
Office	75
Sports centre, no pool	75
Sports centre, with pool	65
Swimming pool	55
Library, museum, gallery	70
Church	90
Hotel	60
Bank, agency	75
Entertainment	75
Prison	60
Court, depot, emergency services building	75
Factory	80

(iii) Once the raw annual space heating energy consumption has been established, it must then be corrected to compensate for variations in climate and exposure by applying the following coefficients:

$$\text{Weather coefficient} = \frac{\text{Standard annual heating degree days}}{\text{Annual heating degree days experienced by building}}$$

In the UK the standard annual number heating degree days is considered to be 2462 (see Appendix 1 for an explanation of degree days). This value will vary with the particular country or region under consideration. Exposure coefficients are as follows:

Exposure	Exposure coefficient
Sheltered (city centre)	1.1
Normal (urban/rural)	1.0
Exposed (coastal/hilly site)	0.9

(iv) The next step is to add the non-heating energy consumption to the corrected space heating energy use to give the raw 'non-time corrected' energy consumption.

(v) The raw annual 'non-time corrected' energy consumption figure should then be multiplied by a coefficient to correct for the 'hours of use' of the building to give the normalized annual energy consumption. This can be done as follows:

$$\text{Hours of use coefficient} = \frac{\text{Standard annual hours of use}}{\text{Actual annual hours of use}}$$

Typical standard annual hours of use values for the UK are shown in Table 4.6.

(vi) The normalized annual energy consumption should be divided by the building floor area to give the NPI. The floor area used in this calculation should exclude completely untreated areas.

(vii) Finally, compare the NPI against the yardsticks given in Table 4.6 and classify the building's energy performance.

Example 4.1 demonstrates the technique involved in calculating a normalized performance indicator for a library building.

Example 4.1

A library building is situated in an urban location, which experiences 2115 heating degree days per year. It is in use for 2400 hours per year, and consumes 940 000 kWh of natural gas and 28 000 kWh of electricity. If the floor area of the school is 4800 m^2, calculate its NPI and assess its energy performance.

Solution

Electrical energy used = 28 000 kWh

Gas used = 940 000 kWh

Total energy consumed = 28 000 + 940 000 = 968 000 kWh

Table 4.5 shows that 70% of the gas used can be attributed directly to space heating.

Table 4.6   Yardsticks (kWh/m² per year) for annual energy consumption of various building types [1]

Building type	Standard hours of use per year	Fair performance range (kWh/m²)
Nursery	2290	370–430
Primary school, no pool	1400	180–240
Primary school with pool	1480	230–310
Secondary school, no pool	1660	190–240
Secondary school, with pool	2000	250–310
Secondary school with sports centre	3690	250–280
Special school, non-residential	1570	250–340
Special school, residential	8760	380–500
Restaurants	–	410–430
Public houses	–	340–470
Fast-food outlets	–	1450–1750
Motorway service area	–	880–1200
Department/chain store (mechanically ventilated)	–	520–620
Other non-food shops*	–	280–320
Superstore/hypermarket (mechanically ventilated)*	–	720–830
Supermarket, no bakery (mechanically ventilated)*	–	1070–1270
Supermarket, with bakery (mechanically ventilated)*	–	1130–1350
Small food shop – general*	–	510–580
Small food shop – fruit & veg	–	400–450
University	4250	325–355
Colleges of Further Education	3200	230–280
Air conditioned offices, over 2000 m²	2600	250–410
Air conditioned offices, under 2000 m²	2400	220–310
Naturally ventilated offices, over 2000 m²	2600	230–290
Naturally ventilated offices, under 2000 m²	2400	200–250
Computer centres	8760	340–480
Swimming pool	4000	1050–1390
Sports centre with pool	5130	570–840
Sports centre, no pool	4910	200–340
Library	2540	200–280
Small hotel	–	240–330
Medium-sized hotel	–	310–420
Large hotel	–	290–420
Banks	2200	180–240
Museum, art gallery	2540	220–310
Cinema	3080	650–780
Theatre	1150	600–900

*Based on sales area

Therefore:

Space heating energy consumption = 940 000 × 0.70 = 658 000 kWh

By applying weather and exposure coefficients:

$$\text{Corrected space heating energy consumption} = 658\,000 \times \frac{2462}{2115} \times 1.0$$

$$= 765\,955.6 \text{ kWh}$$

Non-heating energy consumption = 968 000 − 658 000 = 310 000 kWh

Therefore:

Corrected total energy consumption = 310 000 + 765 955.6 = 1075 955.6 kWh

and correcting for occupancy (using data from Table 4.6):

$$\text{Normalized annual energy consumption} = 1075\,955.6 \times \frac{2540}{2400} = 1138\,719.7 \text{ kWh}$$

Therefore:

$$\text{NPI} = \frac{1138\,719.7}{4800} = 237.2 \text{ kWh/m}^2$$

According to Table 4.6 the assessed energy performance of the library building is 'fair'. In other words, the building is performing reasonably well, but significant energy saving could still be made.

4.4 Time-dependent energy analysis

If enough energy data are collected it is possible to produce a simple graph in which energy consumption is plotted against time (see Figure 4.2). Through this type of simple time-dependent analysis it is possible to identify general trends and seasonal patterns in energy consumption. This can prove invaluable, since it enables exceptions to the norm to be identified quickly. Although a useful tool, it is important to understand the limitations of this technique, which is best used as a comparative tool and not an absolute one.

Consider the case of the office building represented in Figure 4.2. The time-dependent graph shows monthly gas consumption for the years 1994 and 1995. It can be seen from the graph that:

- Energy consumption during the months of January, February and March of 1995 is consistently less than in the corresponding period in 1994.

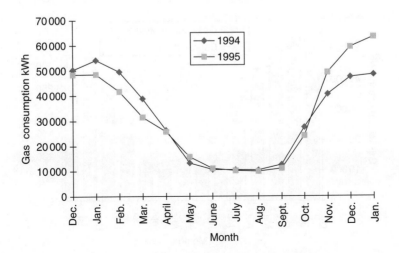

Figure 4.2   Office building gas consumption

- The base load consumption is approximately 10 500 kWh/month. This presumably is the gas consumed in producing domestic hot water and in catering.
- Energy consumption during the months of November and December 1995, and January 1996 appears to have increased significantly compared with the corresponding 1994 figures. This tends to indicate a loss of control in the heating system, which might have arisen as a result of an operative altering the control settings at the end of October.

Although energy consumption in January, February and March of 1995 is consistently lower than that in the corresponding period in 1994, it should be noted that this could be for a variety of reasons, some of which are as follows:

- Improved operating practices.
- Warmer weather in 1995, compared with 1994.
- Less hours worked by staff in the office. Perhaps during this period in 1994 the office was open during the weekend, because of a high workload!

It is impossible to identify precisely from Figure 4.2 why the energy consumption for January, February and March 1995 is lower than for the same period in 1994. In order to do this more sophisticated analytical techniques are required (see Section 4.5). Notwithstanding this, it is possible using a time related graph, to plot more than one variable against time, as in Figure 4.3 where the gas energy consumption figures for 1994 are plotted along side the relevant degree day data.

By generating time-related graphs such as Figure 4.3 it is possible to:

- Identify cyclical patterns which indicate seasonal loads. As the monthly degree day total rises, so gas consumption should also rise.
- Identify general trends which reflect changes in energy consumption. These may arise because of changes in load or efficiency, or alternatively they may be due to changes in operating practice.
- Identify a steady base load, which is the energy consumed when plant is operating at minimum load. For example, in Figure 4.3 a base load of approximately 10 500 kWh per month occurs

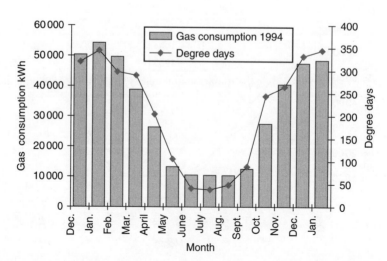

Figure 4.3 Comparison of gas consumption and degree days for 1994

in June, July and August when no space heating is required. This base load probably represents domestic hot water production and catering gas consumption.

- Identify a lack of any clear pattern. This usually represents a system which is suffering from lack of good control.
- Identify periods of very high or very low energy consumption, which may occur because of unusual changes in plant operation in a particular month. It should always be remembered that anomalies might appear because of errors in logging energy input data.

Most building energy management applications use monthly data. It is therefore recommended that data be based on calendar months, since analysis of data derived from 4- or 5-week periods might cause errors. When using calendar months, the one or two day differences, which occur between successive months, can usually be ignored.

4.5 Linear regression analysis

Linear regression analysis is a statistical technique which determines and quantifies the relationship between variables. It is a widely used energy management tool which enables standard equations to be established for energy consumption, often from data which would otherwise be meaningless.

From Section 4.4 it is clear that although time-dependent analysis is a useful comparative tool, it has its limitations; it is difficult to identify why certain trends occur or indeed, if perceived trends actually exist at all. Regression analysis overcomes this problem by removing the 'time' element from the analysis and focusing instead on the variables which influence energy consumption. It is a versatile technique which can be used to analyse a wide variety of applications. When used as an energy management tool, the variables commonly compared are [2]:

- Gas consumption versus the number of heating degree days experienced.
- Gas consumption versus the number of units of production.
- Electricity consumption versus the number of units of production.
- Water consumption versus the number of units of production.
- Electricity consumed by lighting versus hours of occupancy.

Regression analysis is very much dependent on the quality of the data used. It should therefore be treated with care. If an analysis indicates the absence of a significant relationship (i.e. $P < 0.05$) between two variables, it does not necessarily mean that no relationship exists. The significance of results depends on the quantity and quality of the data used, and indeed, on the variables used in the analysis. Table 4.7 shows a selection of variables which can influence energy and water consumption.

4.5.1 Single independent variable

Consider a case where the monthly gas consumption of an office building (i.e. a dependent variable) and the number of heating degree days experienced (i.e. an independent variable) are plotted against each other on a graph. Since it is well known that building heat losses increase as the outside air temperature gets colder, it is reasonable to expect some sort of relationship between the two. This relationship is in fact linear and it is possible to derive an equation for the *best-fit* straight-line curve through the points plotted on the graph. The *best-fit* straight-line curve is determined by summing the squares of the distances from the straight line of the various data

Table 4.7   Factors which influence energy consumption [2]

Commodity	Duty performed	Possible factors
Electricity	Outside security lighting	Hours of darkness
Water	Swimming pool make-up	Number of bathers (because of evaporation and water removed in swimming costumes)
Gas	Space heating	Heating degree days
Electricity	Air conditioning	Cooling degree days
Oil	Steam-raising in boiler plant	Amount of steam generated
Electricity	Air compressor	Air volume delivered
Diesel road	Goods vehicles	Tonne-miles hauled
Fuel steam	Production process	Production volume

points. Once established, this linear equation can be used to predict future energy consumption. In addition, it can be used as a *standard performance* equation for energy *monitoring and targeting* purposes (see Chapter 7).

The generic equation for a straight-line graph can be represented as:

$$y = c + mx \tag{4.1}$$

Where y is the dependent variable (e.g. energy consumption), x is the independent variable (e.g. number of degree days), c is the value at which the straight-line curve intersects the 'y' axis, and m is the gradient of the straight-line curve.

If the straight line $y = c + mx$ is best fitted to a set of data sample points;

$$(x_1, y_1) \cdot (x_2, y_2) \cdots (x_n, y_n)$$

it can be shown that

$$cn + m\Sigma x = \Sigma y \tag{4.2}$$

and

$$c\Sigma x + m\Sigma x^2 = \Sigma xy \tag{4.3}$$

Where n is the number of data points.

These equations are known as the normal equations of the problem and they can be used to establish the values of c and m, as illustrated in Example 4.2.

Example 4.2

Consider a hospital building which during a monitoring programme produces the following sample data:

Degree days experienced per month (x)	72	88	95	106	169	204	244	265	290	298	332	345
Gas consumption per month (y) (GJ)	482	520	634	570	671	860	903	940	1007	1210	1020	1131

Therefore:

x	y	x^2	xy
72	482	5184	34 704
88	520	7744	45 760
95	634	9025	60 230
106	570	11 236	60 420
169	671	28 561	113 399
204	860	41 616	175 440
244	903	59 536	220 332
265	940	70 225	249 100
290	1007	84 100	292 030
298	1210	88 804	360 580
332	1020	110 224	338 640
345	1131	119 025	390 195
Σ 2508	9948	635 280	2340 830

Therefore, the normal equations become:

$$12c + 2508m = 9948$$

and

$$2508c + 635\ 280m = 2340\ 830$$

therefore

$$c = \frac{9948 - 2508m}{12}$$

therefore

$$2508\,\frac{(9948 - 2508m)}{12} + 635\ 280m = 2340\ 830$$

therefore

$$m = 2.355$$

and

$$c = 336.73$$

The best fit straight line curve equation is therefore:

$$y = 336.73 + 2.355x$$

From this equation, it can be seen that the theoretical base load for the building is 336.73 GJ. The graph resulting from the regression analysis is shown in Figure 4.4.

The linear regression curve in Figure 4.4 shows that even when zero degree days are experienced in a particular month, the building still consumes 336.7 GJ of gas. This implies that the theoretical monthly base load gas consumption for catering and hot water production is 336.7 GJ, and that the annual base load consumption is about 4040.4 GJ (i.e. 12 × 336.7 GJ).

4.5.2 Correlation coefficients

The regression analysis method described in Section 4.5.1 enables a *best-fit* straight line to be determined for a sample data set. However, in some circumstances the sample data points may

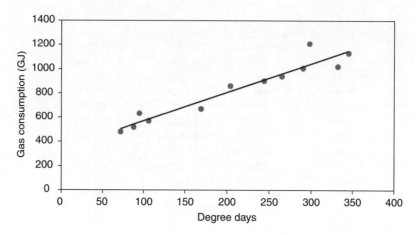

Figure 4.4   Regression analysis for hospital building

be very scattered with the result that the derived equation may be meaningless. It is therefore important to determine how well the *best-fit* line correlates to the sample data. This can be done by calculating the Pearson correlation coefficient [3], which gives an indication of the reliability of the line drawn. The Pearson correlation coefficient is a value between 1 and 0, with a value of 1 representing 100% correlation. The Pearson correlation coefficient (r) can be determined using eqn (4.4).

$$r = \frac{\Sigma(x - \bar{x})(y - \bar{y})}{\sqrt{[\Sigma(x - \bar{x})^2 \Sigma(y - \bar{y})^2]}}$$

(4.4)

Where x, y are the x and y values, and $\bar{x}$, $\bar{y}$ are the average x and y values.
Example 4.3 illustrates how the correlation coefficient may be calculated.

Example 4.3

For the data presented in Example 4.2 determine the correlation coefficient.
Therefore:

	x	y	$(x - \bar{x})$	$(y - \bar{y})$	$(x - \bar{x})(y - \bar{y})$	$(x - \bar{x})^2$	$(y - \bar{y})^2$
	72	482	−137	−347	47 539	18 769	120 409
	88	520	−121	−309	37 389	14 641	95 481
	95	634	−114	−195	22 230	12 996	38 025
	106	570	−103	−259	26 677	10 609	67 081
	169	671	−40	−158	6 320	1 600	24 964
	204	860	−5	31	−155	25	961
	244	903	35	74	2 590	1 225	5 476
	265	940	56	111	6 216	3 136	12 321
	290	1007	81	178	14 418	6 561	31 684
	298	1210	89	381	33 909	7 921	145 161
	332	1020	123	191	23 493	15 129	36 481
	345	1131	136	302	41 072	18 496	91 204
Σ	2508	9948	0	0	261 698	111 108	669 248

Therefore:

$$r = \frac{261\ 698}{\sqrt{(111\ 108 \times 669\ 248)}} = 0.96$$

Table 4.8 shows minimum acceptable correlation coefficients for given numbers of data samples.

It can be seen from Table 4.8 that the correlation coefficient in Example 4.3 is very good.

Table 4.8   Minimum correlation coefficients [4]

Number of data samples	Minimum correlation coefficient
10	0.767
15	0.641
20	0.561
25	0.506
30	0.464
35	0.425
40	0.402
45	0.380
50	0.362

4.5.3 Multi-variable analysis

Often energy consumption can be influenced by several different variables. When this is the case the relationship can be described by the equation:

$$y = c + m_1x_1 + m_2x_2 + \cdots + m_nx_n \tag{4.5}$$

where $x_1, x_2, \ldots, x_n$ are the variables that influence y.
Examples of where multiple variables influence energy consumption could be:

- A factory building where electricity consumption is influenced by both the volume of production and the hours of darkness experienced; or
- An air conditioning office building where electricity consumption is influenced by both the cooling degree days and the hours of darkness experienced.

It is difficult to solve multivariable analysis by hand calculation. It is therefore advisable to use specialist computer software which can be employed to determine the statistical relationship between the variables.

4.6 CUSUM

Regression analysis enables the relationship between energy use and variables such as heating degree days to be established for a given period. It can be used to establish a base line *standard performance* equation, against which subsequent energy consumption can be measured. One technique that can be employed to assess subsequent energy consumption is known as CUSUM, which is an acronym for *cumulative sum deviation method* [2, 5]. It is a measure of the progressive deviation from a standard consumption pattern. It is simple to calculate and involves the cumulative summation of the differences between actual energy consumption and target, or base line, energy consumption. Base line values should be calculated from a *standard performance* equation which should be derived through analysis of data collected during a monitoring period before any interventions are made.

In order to produce a CUSUM plot the following steps should be taken:

(i) Plot a scatter graph of the two variables under consideration (similar to that shown in Figure 4.4) for the 'base line' period and derive a *standard performance* equation as described in Section 4.5.1.

(ii) Use the *standard performance* equation to calculate the predicted energy consumption for each month (including the period covered by the baseline and any subsequent study months).

(iii) For each data point (i.e. for each month), subtract the predicted consumption from the actual consumption.

(iv) For each data point, obtain the cumulative total deviation from predicted consumption; this gives the CUSUM value for each data point.

(v) Plot the CUSUM values against time.

The CUSUM process is illustrated by Example 4.4, which shows the gas consumption for an office building over a 44 month period.

Example 4.4

An energy audit of an office building for the period August 1989 to December 1990 produced the following gas consumption data:

Year	Month	Gas consumed (kWh)	Heating degree days
1989	August	15 490	18
	September	23 700	36
	October	55 673	109
	November	94 382	199
	December	106 683	239
1990	January	110 745	247
	February	96 458	210
	March	95 903	207
	April	93 265	195
	May	60 045	117
	June	32 267	58
	July	18 849	24
	August	12 435	12
	September	32 775	60
	October	43 924	95
	November	95 012	201
	December	129 505	280

A subsequent monitoring programme found gas consumption for the period January 1991 to March 1993 to be as follows:

Year	Month	Gas consumed (kWh)	Heating degree days
1991	January	140 022	308
	February	180 034	338
	March	118 524	214
	April	112 045	201
	May	64 045	108
	June	37 724	67
	July	18 490	24
	August	17 045	21
	September	22 483	35
	October	66 275	140
	November	101 040	219
	December	120 500	262
1992	January	144 240	323
	February	123 140	271
	March	91 500	232
	April	78 041	195
	May	41 004	96
	June	20 549	39
	July	13 461	18
	August	16 062	26
	September	28 740	61
	October	44 467	103
	November	77 206	197
	December	112 442	290
1993	January	98 950	260
	February	98 399	253
	March	97 760	250

Given the information above produce a CUSUM plot for the building.

Solution

Using the linear regression technique described in Section 4.5.1 the *standard performance* equation for the base line period is:

$$\text{Monthly gas consumption} = 7744.7 + (427.16 \times \text{degree days})$$

Once this is established it is possible to produce the CUSUM results, as shown in Table 4.9. From the results in Table 4.9 it is possible to produce the CUSUM plot, as shown in Figure 4.5.

A cursory inspection of the CUSUM graph in Figure 4.5 reveals that two major events occurred during the monitoring period, one in February 1991 when the energy consumption dramatically increased, and the other in February 1992 when it started to decrease. In fact further investigation revealed that:

- In February 1991 the time clock on the heating system was incorrectly set, so that the heating remained on until 22.00 hours, leading to a dramatic increase in gas consumption. This problem was spotted and rectified in May 1991, at which point the CUSUM plot returns to a similar gradient as that experienced during the base line period.

Table 4.9   CUSUM calculation

Year	Month	Gas consumed (kWh) (1)	Heating degree days (2)	Target gas used (kWh) (3)	Difference (1–3) (kWh) (4)	CUSUM (kWh) (5)
1989	August	15 490	18	15 434	56	56
	September	23 700	36	23 122	578	634
	October	55 673	109	54 305	1 368	2 002
	November	94 382	199	92 749	1 633	3 635
	December	106 683	239	109 835	−3 152	483
1990	January	110 745	247	113 252	−2 507	−2 024
	February	96 458	210	97 448	−990	−3 014
	March	95 903	207	96 166	−263	−3 277
	April	93 265	195	91 040	225	−1 052
	May	60 045	117	57 722	2 323	1 270
	June	32 267	58	32 520	−253	1 018
	July	18 849	24	17 996	853	1 870
	August	12 435	12	12 871	−436	1 435
	September	32 775	60	33 374	−599	836
	October	43 924	95	48 325	−4 401	−3 565
	November	95 012	201	93 603	1 409	−2 156
	December	129 505	280	127 349	2 156	0
1991	January	140 022	308	139 309	713	713
	February	180 034	338	152 124	27 910	28 623
	March	118 524	214	99 156	19 368	47 991
	April	112 045	201	93 603	18 442	66 433
	May	64 045	108	53 878	10 167	76 600
	June	37 724	67	36 364	1 360	77 960
	July	18 490	24	17 996	494	78 453
	August	17 045	21	16 715	330	78 783
	September	22 483	35	22 695	−212	78 571
	October	66 275	140	67 547	− 1 272	77 299
	November	101 040	219	101 292	−252	77 047
	December	120 500	262	119 660	840	77 888
1992	January	144 240	323	145 716	−1 476	76 411
	February	123 140	271	123 504	−364	76 047
	March	91 500	232	106 845	−15 345	60 702
	April	78 041	195	91 040	−12 999	47 703
	May	41 004	96	48 752	−7 748	39 955
	June	20 549	39	24 404	−3 855	36 100
	July	13 461	18	15 434	−1 973	34 127
	August	16 062	26	18 851	−2 789	31 339
	September	28 740	61	33 801	−5 061	26 277
	October	44 467	103	51 742	−7 275	19 002
	November	77 206	197	91 895	−14 689	4 314
	December	112 442	290	131 620	−19 178	−14 864
1993	January	98 950	260	118 806	−19 856	−34 720
	February	98 399	253	115 815	−17 416	−52 136
	March	97 760	250	114 534	−16 774	−68 910

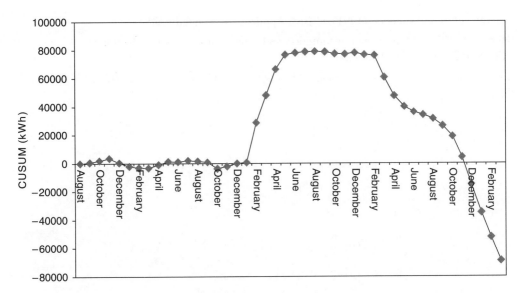

Figure 4.5   CUSUM graph

- In February 1992 the old single glazing in the office building was replaced by double glazed window, thus leading to a fall in energy consumption.

It should also be noted that:

- During the summer months the slope of the CUSUM line became less steep, due to the reduced demand for heating.
- The base line should always be a horizontal line through zero. This is because by definition it is a *best-fit* straight line for the data samples occurring during the base line period. The actual samples during this period may deviate from the predicted base line figures, but the CUSUM plot will always fluctuate about zero.
- The greater the slope downwards of the CUSUM line, the more energy efficient the process.

References

1. Energy audits and surveys. (1991). CIBSE Applications Manual AM5.
2. Waste avoidance methods. (1995). Fuel Efficiency Booklet 13, Department of the Environment.
3. Campbell, M. J. and Machin, D. (1999). *Medical statistics: a commonsense approach*. Appendix 1. John Wiley and Sons.
4. Measurement of energy consumption: CIBSE Building Energy Code Part 4. (1982). CIBSE.
5. Harris, P. (1985). Monitoring consumption and setting targets. *Energy Manager's Workbook*, **2** (Chapter 2), Energy Publications.

Bibliography

Contract energy management. (1991). CIBSE Applications Manual AM6.

Eastop, T. D. and Croft, D. R. (1990). *Energy efficiency for engineers and technologists* (Chapters 7 and 9). Longman Scientific & Technical.

Energy audits and surveys. (1991). CIBSE Applications Manual AM5.

Harris, P. (1985). Monitoring consumption and setting targets. *Energy Manager's Workbook*, **2** (Chapter 2), Energy Publications.

Moss, K. J. (1997). *Energy Management and Operating Costs in Buildings* (Chapter 10). E&FN Spon.

5

Energy audits and surveys

Before any energy saving measures can be undertaken within an organization, it is first necessary to collect comprehensive energy data through an auditing process. This chapter focuses on the mechanics of energy auditing and describes the procedures involved. The differences between preliminary, targeted and comprehensive audits are discussed and the procedures associated with each type highlighted. The subject of energy surveys is also discussed.

5.1 Introduction

There is a strong analogy between the medical profession and the field of energy management. If a patient with a medical complaint presents him or herself before a doctor, the doctor must first accurately diagnose the condition before taking further steps. The doctor should obtain information from the patient by asking informed questions, possibly carrying out tests, use knowledge and expertize in order to diagnose the complaint and ultimately prescribe treatment. In a similar manner, before any energy 'problems' can be treated it is first necessary to determine the current state of a facility's, or organization's energy consumption and thus diagnose any problems that exist. In order to do this, an energy audit must be undertaken and analysis performed on the data collected.

An energy audit is a feasibility study to establish and quantify the cost of the various energy inputs to, and flows within, a facility or organization over a given period. The overall aim of an energy audit is to identify viable and cost effective energy measures which will reduce operating costs. Energy audits can take a variety of forms, but the process usually involves collecting data from energy invoices and meters, and undertaking surveys of plant, equipment and buildings, as well as collecting information from managers and other staff. The auditing process should identify ways to enhance an organization's operating efficiency and decrease its maintenance costs. In addition, the process should help to resolve any occupant comfort problem which may exist.

An energy audit should be viewed as the 'foundation' on which any future energy management programme is built. Energy management programmes (discussed in detail in Chapter 7) involve the continual monitoring and targeting of energy consumption. Before targets can be set, or effective monitoring undertaken, it is important to establish:

- Baseline energy consumption;
- Patterns of operation and the work practices used;
- The condition of the organization's buildings, plant and equipment; and
- *Energy management opportunities*, which will result in energy cost reductions.

This information can only be obtained by carrying out a full energy audit of an organization's facilities.

An energy audit should identify those issues which need immediate direct action, as well as those which require further detailed investigation. It should also produce data which can be used to justify future capital investment, and raise, within the organization, general awareness of energy conservation matters. The financial benefits afforded to an organization by an energy audit are both direct and indirect. The direct benefits are fairly obvious; energy cost savings can be achieved by reducing consumption, or simply by changing tariff or fuel type. The indirect benefits are much less obvious; reduced maintenance costs will arise from improved plant utilization and reduced operating hours. Also, improved plant utilization may result in the elimination of excess plant capacity and ultimately reduce capital expenditure.

The auditing process should identify *energy management opportunities*, which when implemented will result in financial benefit to an organization. The magnitude of these financial benefits is not necessarily dependent on the level of capital investment. In many situations, major cost savings can be achieved through the implementation of 'no cost' or low cost measures, such as:

- Changing an energy tariff;
- Rescheduling production activities to take advantage of preferential tariffs;
- Adjusting existing controls so that plant operation matches the actual requirements of the building or manufacturing process;
- Implementing good housekeeping policies, in which staff are encouraged to avoid energy wasteful practices;
- Investment in small capital items such as thermostats and time switches.

Although much can be achieved through low cost measures, it is sometimes necessary to undertake more capital-intensive measures, such as replacing worn out plant or installing a building management system (BMS). Because of the capital involved in such measures, decisions to invest in them are usually made by senior management (see Chapter 6). In this situation the results of an energy audit can be used to justify capital investment.

5.2 Types of energy audit

Although there are many variations, energy audits can broadly be classified as *preliminary, targeted and comprehensive* audits. Each type is distinguished by the level of detail involved and the depth of the analysis undertaken. It is important to select the appropriate audit type for the facility concerned. Comprehensive audits involve detailed energy surveys of plant, equipment and the fabric of buildings, which is a time consuming and expensive process. They therefore should not be undertaken lightly. It is often better to focus detailed surveys on problem areas highlighted by a preliminary energy audit, otherwise much time and money can be wasted. By carrying out a preliminary audit and methodically applying a range of simple analysis techniques it is often possible to identify major energy problems without the need for expensive and detailed energy surveys.

Preliminary energy audits seek to establish the quantity and cost of each form of energy used in a facility or in an organization. They are relatively quick and are designed to determine a project's potential; more detailed energy audits and surveys can always be undertaken later if so required. Preliminary audits are primarily concerned with obtaining data from energy invoices and meter readings for a given period, which often corresponds to the most recent financial year. Because such audits are primarily concerned with gathering data from bills and invoices, it is sometimes helpful to think of preliminary audits as being *financial energy audits.*

Targeted energy audits often result from preliminary audits. They provide data and detailed analysis on specific targeted projects. For example, an organization may target its lighting installation or boilers with a view to up-grading these items of equipment. Targeted audits therefore involve detailed surveys of the target subjects and analysis of the energy flows and costs associated with the targets. They should make recommendations regarding action to be taken.

Comprehensive energy audits provide detailed data on the energy inputs to, and energy flows within, a facility or organization. They should produce detailed energy project implementation plans. Such audits involve detailed energy surveys and may involve the use of complex energy simulation computer software.

Although in-house staff can carry out audits, they are generally undertaken by either specialist energy consultants or energy service companies. Energy service companies make their money through performance contracts, which guarantee organizations energy cost savings in return for negotiated fees. To energy service companies the main interest lies not in the audit itself, but in installing and managing plant in accordance with their recommendations. Some companies may even arrange the finance for such projects. When using an energy service company it is thus important to remember that they have a vested interest in the outcome of any energy audit and that they may not be totally impartial. By contrast, energy consultants are independent and therefore should provide objective advice.

5.2.1 Audit costs

Energy audits can be expensive undertakings. Table 5.1 shows data produced by the California Energy Commission for the cost of performing an energy audit in the USA [1].

It can be seen from Table 5.1 that the more complex the audit, the higher the costs involved. It is therefore important to select the appropriate level of audit for any particular application.

Audit costs are affected by the complexity of the facility under consideration. For example, complex facilities such as hospitals or universities are more costly to audit than say, schools. The age of the facility may also affect the cost. For example, if a mechanical system is complex and the 'as built' drawings are out of date or not available then the energy auditors may have to produce schematic drawings. This can be very time consuming and obviously greatly increases

Table 5.1   Energy audit costs in the USA (1997 rates) [1]

Type of energy audit	Typical cost ($/sq. ft.) (1997 dollar rate)
Preliminary audit	$0.144–$0.333 per m^2
Targeted audit	$0.333–$0.778 per m^2 (lighting projects) $0.556–$1.00 per m^2 (HVAC and controls projects)
Comprehensive audit	$2.00–$5.556 per m^2 (less than 4500 m^2) Less than $1.333 per m^2 (more than 22500 m^2)

the audit costs. Given the cost involved, it is important that organizations assist their auditors by preparing in advance for the audit and providing the auditing team with as much relevant information as possible. Energy bills, fuel invoices, meter readings and operational notes should all be collected, together with any relevant system or building drawings. Organizations should also inform their management team that an energy audit is being undertaken and arrange for the auditors to meet with key managers and other relevant staff.

5.3 Why is energy wasted?

Before looking in detail at the processes involved in energy auditing, it is perhaps worth looking briefly at the reasons why energy is wasted in so many organizations. Energy is often wasted because of:

- Poorly designed buildings and installations. Buildings may be poorly insulated resulting in high space heating costs, or mechanical ventilation ducts may be undersized so that fan power consumption is high.
- Inadequate control systems. Heating systems may be installed without any optimum start control.
- Poor control settings. Time clock controllers may be incorrectly set so that buildings are heated when not in use.
- Inefficient plant operation, often arising from the use of old or out of date technology, a situation often made worse by poor maintenance practices.
- Poor operating and working practices. Lights are often left on in buildings when they should be switched off.

Although the reasons for energy waste are multifactorial, some of the main reasons are as follows:

- Building designers do not pay energy bills. The design process is closely allied to the construction process, and designers usually select low capital cost solutions, which often result in higher operating costs. This situation is made worse by the fact that the budgets for constructing a facility and running it are usually completely separate.
- Energy consumption is taken for granted. Most building occupants and users do not pay energy bills. They are concerned with their own personal comfort and are not particularly interested in how much energy is consumed in achieving a comfortable environment.
- Most organizations do not have a culture of energy efficiency.
- In many countries the cost of energy is low in comparison with labour costs.

The above list demonstrates that much energy wastage arises from poor strategic and operational management, and also a lack of an energy saving culture amongst staff. Energy can often be saved at no capital expense simply by improving maintenance procedures and instigating good work practices. This is often referred to as 'good housekeeping', and involves simple measures such as encouraging personnel to switch off lights when they are not required. Initiating good maintenance procedures is also important. For example, if filters in ducted air handling systems are not replaced regularly, then they become dirty with the result that fan energy consumption increases. It has been estimated that energy bills for organizations can be reduced by approximately 20% through the use of good energy management practices [2]. It is therefore important that the human and management aspects of energy consumption are investigated in any energy audit. Without a supportive management culture it is difficult to make lasting energy savings in any organization.

5.4 Preliminary energy audits

Preliminary audits seek to quantify and cost each form of energy input to a facility or organization over a period of time. They should also identify where the energy is being used within the
organization. The main processes involved in such an audit are:

- Collecting data
- Analysing data
- Presenting data
- Establishing priorities and making recommendations.

At the start of any audit process it is important to gather preliminary data about the geographic
location of the particular facility concerned, together with any relevant distinguishing features
such as its altitude and orientation. Local weather data and degree day data covering the audit
period should also be collected. These data will act as a benchmark reference against which the
facility's energy consumption can be measured. With manufacturing facilities it will also be
necessary to collect data concerning the production output during the audit period, since this will
have a considerable impact on energy consumption.

Probably the single most important source of energy data is the energy invoice. It is therefore
very important that the audit team has all the relevant energy invoices for the selected audit
period. By compiling data from invoices it is possible to build up a clear picture of the pattern
of energy consumption and the associated costs to a facility of the various energy inputs. In
addition, the total amount spent on energy can be determined from the invoices, thus indicating
the upper limit which can ultimately be saved through energy management measures.

When collecting data from fuel and energy invoices it is important to ensure that copies of all
utility invoices for the audit period are collected rather than simply those for which payments
were made during the audit period. It is also important to collect all the invoices or delivery notes
relating to oil, solid fuel or liquid petroleum gas for the audit period. Due to the time lag between
delivery and consumption, it may also be necessary to include deliveries which occurred before
the start of the audit period. In addition, it is essential that all the metering and supply points are
identified from the invoices, to account for all the energy inputs.

Any estimated meter readings should be identified, since these can result in misleading data.
In order to overcome the problems associated with estimated readings, additional invoices
should be collected which cover the same months as the estimated invoice, but for years prior to
the audit period. These 'real' data can then be compared with the estimated data, to establish
realistic data for the audit period. Where possible, data from invoices should be corroborated by
independent meter reading data collected over the audit period.

If invoice data are inadequate or unavailable, then it will be necessary to approach the utility
companies or fuel suppliers for assistance.

Although for most facilities it will be relatively simple to identify the utility metering points,
on large complex sites it may be difficult to account for all the meters. Utility services may come
from a variety of sources and this will be reflected in the invoices. Consider a large supermarket
site which has an 11 kV electricity supply to the main shopping complex and a separate incoming
415 V supply to a remote petrol station owned by the retailer on the same site. Both supplies will
be billed separately, and may be on different tariffs. The two supplies might even come from
different utility companies!

Preliminary analysis of energy invoices can often be very useful in identifying any anomalies
which require further investigation. If a relatively small building on a site consumes as much
gas as one of its much larger neighbours, then it would appear that something is wrong.

Further investigation can then be undertaken, which might reveal that the high gas consumption is due to the heating plant in the small building operating at nighttime when the building is empty.

Given that the gathering of information from invoices is crucial to the auditing process, it is important that energy invoices be understood. It is also important to understand the peculiarities of the various fuel types and utility services, since these can have a bearing on energy consumption and on the auditing process itself.

5.4.1 Electricity invoices

The precise nature of electricity tariffs and supply contracts is discussed in detail in Chapter 2. For the purpose of this chapter we will consider only monthly maximum demand tariffs, since many medium-sized and large organizations use tariffs of a similar nature. Figure 5.1 shows a simplified electricity invoice, which illustrates many of the features commonly found in monthly electricity bills.

The monthly electricity invoice shown in Figure 5.1 contains the following information:

- The date of the meter reading.
- The monthly standing charge, which is £30.00 in Figure 5.1.
- The present and previous meter readings with the number of units supplied. These are usually divided into two sets of meter readings, daytime units (i.e. peak rate) and nighttime units (i.e. off-peak rate). The difference between the present and previous readings is the units of electricity consumed in the period since the previous meter reading. With some meters a constant may be included on the invoice; multiplication of the meter advances by this constant gives the actual number of units supplied in kVAh or kWh.
- The charges for each unit of electrical energy consumed. These are usually different for the daytime and nighttime. In Figure 5.1 the peak rate is 6.05 p per kVAh and the off-peak rate is 2.60 p per kVAh. Sometimes blocks of units are charged at different rates. Some tariffs levy a higher unit charge on the first 1000 kVAh or kWh consumed.

METER READINGS			METER READING DATE: 23/01/2000	
Present	Previous		Units Consumed	
247451	224520		22931	
184530 (Night)	174702 (Night)		9828 (Night)	
MAXIMUM DEMAND THIS MONTH	270 kVA		**ANNUAL MAXIMUM DEMAND**	300 kVA

DESCRIPTION OF CHARGE	No. OF UNITS OR kVA	RATE	AMOUNT EXCLUSIVE OF TAX	TAX (VAT)	% RATE
MONTHLY CHARGE		£30.00	30.00	5.25	17.50
AVAILABILITY CHARGE	300.0	£1.41	423.00	74.03	17.50
MAX.DEMAND CHARGE	270.0	£5.35	1444.50	252.79	17.50
UNIT CHARGE	22931	6.05p	1387.32	242.78	17.50
NIGHT UNITS	9828	2.60p	255.53	44.72	17.50
		TOTAL	3540.35	619.57	
	TOTAL DUE	£4159.92			

Figure 5.1   Monthly electricity bill

- A monthly maximum demand charge for every kW or kVA of the peak power demand occurring during the billing month. Maximum demand charges are designed to penalize users who make heavy demands on the supply grid during peak periods. They vary throughout the year; in northern countries such as the UK, they are at their highest in December and January and very low, or non-existent, during the summer. In hot countries where air conditioning is extensively used, the situation is reversed with the highest demand charges being levied in summer. Demand charges are often stepped and levied at different rates for various parts of the year. Figure 5.1 shows a maximum demand of 270 kVA, all of which is charged at £5.35 per kVA.
- The supply capacity, for which a monthly availability charge is levied. In Figure 5.1 chargeable supply capacity is 300 kVA, all of which is charged at £1.41 per kVA each month.
- The VAT charged on the bill, together with the total cost due.

5.4.2 Natural gas

Natural gas invoices are generally much less complicated than their electricity counterparts. Figure 5.2 shows a typical example of a monthly gas invoice.

The gas invoices in Figure 5.2 includes the following:

- The date of meter reading or estimate.
- The calorific value of gas (i.e. 39.6 MJ/m^3 in Figure 5.2);
- The present and previous readings with the amount of gas used, often presented in cubic feet, cubic metres or kWh (and sometimes in therms);
- The unit price per kWh of natural gas (i.e. 1.520 p/kWh in Figure 5.2);
- A fixed monthly or quarterly standing charge (i.e. £9.45 per kWh in Figure 5.2).
- The VAT charged on the bill, together with the total cost due.

The unit price of natural gas may be fixed as in Figure 5.2, or it may vary depending on the volume of gas consumed. In addition, some large sites may have a combination of firm and interruptible supplies, which may be invoiced separately.

5.4.3 Fuel oil

Fuel oil is measured by volume, which varies with temperature. Delivery invoices for fuel oils should therefore state the volume corrected to a standard condition of 15.5 °C [3]. They should also state the date of delivery, the delivery note reference number, the unit cost per standard litre,

DATE OF BILL	03/04/2000	DATE OF METER READING	16/03/2000	CALORIFIC VALUE (MJ/m^3)	39.6

METER READING		GAS USED			COSTS
Present	Previous	Cubic Feet	Cubic Metres	kWh	
3171	2825	346	979.1	10770	163.70

STANDING CHARGE					9.45
£173.15 AT 17.5% VAT					30.30
		UNIT CHARGE	1.520p/kWh		
				TOTAL	£203.45

Figure 5.2   Typical gas bill

and the VAT charged on the bill. The calorific value of the oil may be included, but if not, this should be obtained from the supplier. This last point is important because the calorific value varies with the type of fuel oil used.

5.4.4 Solid fuel

Solid fuel invoices generally state the weight delivered and the cost, but do not always include data on the calorific value. These data can usually be obtained from the supplier. Solid fuels can present particular problems to energy auditors because they are often stock piled, making accurate short-term assessment of solid fuel consumption a difficult task.

5.4.5 Heat

In many parts of northern Europe buildings and whole towns rely on heat produced in cogeneration plants. The heat is usually supplied in the form of medium-pressure or high-pressure water from a district heating main and transferred to individual buildings via heat exchangers. The heat energy consumption is recorded by heat meters, which record the water flow rate, and the temperatures of the water entering and leaving the facility, thus determining the energy consumed in kWh. It is important to note that the accuracy of heat meters can be affected by variations in temperature and flow rate. At low flow rates or where small temperature differences occur, metering errors can be significant.

5.4.6 Site records

In larger more complex facilities, especially those which employ an energy manager, it is often the case that site energy records are kept. These can be an important source of information to an audit team and can be used to corroborate data collected from energy invoices. In particular, records of sub-meter readings can be particularly useful, since they give detailed information about energy flows. However, when using data collected from sub-meters, it is important to know where the meters are located and to understand what they are measuring. It should be realized that sub-meters are always 'subordinate' to main meters. In other words, the energy consumption recorded by a sub-meter is always a sub-set of that recorded by the main meter and not additional energy consumption. Failure to recognize this will result in major errors. One good way of avoiding these errors is to construct schematic diagrams showing the respective positions of all the main meters and sub-meters. In addition, it should be remembered that meter records often contain mistakes (see Chapter 7). Care should therefore be taken to validate meter readings.

5.4.7 Data analysis

The data analysis techniques used in energy audits are described in detail in Chapter 4. For a preliminary energy audit, analysis should be limited to those techniques which enable the auditor to determine:

- How much energy is being consumed;
- What type of energy is being consumed;
- The performance of the facility compared with other similar facilities; and
- The characteristic performance of the building.

These outcomes can be achieved by using the percentage breakdown technique (described in Section 4.2), the NPI technique (described in Section 4.3) and the linear regression technique (described in Section 4.5).

When all the energy data have been gathered and analysed, they need to be compared against various 'yardsticks' for similar facilities. Table 5.2 gives the NPIs for various building types in the UK, together with percentage breakdowns for typical energy consumption [3]. Example 5.1 illustrates how the analysis for a preliminary energy audit might be performed.

Example 5.1

A preliminary energy audit of a 5000 m^2 air conditioned office building has yielded the following energy data:

Month	Heating degree days	Gas consumption (kWh)	Gas cost (£)	Electricity consumption (kWh)	Electricity cost (£)
January	267	90 010	1080.12	68 214	3956.41
February	298	97 160	1165.92	60 312	3437.78
March	250	87 058	1044.70	59 645	3280.48
April	176	71 320	855.84	65 045	3382.34
May	69	47 200	566.40	89 234	4550.93
June	30	38 645	463.74	105 932	5296.60
July	12	33 840	406.08	119 237	5961.85
August	20	34 400	412.80	103 247	5265.60
September	50	44 050	528.60	88 235	4588.22
October	208	75 920	911.04	65 023	3446.22
November	215	78 580	942.96	61 567	3447.75
December	337	106 640	1279.68	70 124	4137.32
Totals	1932	804 823	9657.88	955 815	50 751.50

Given that the office building is located in a city centre and is occupied for 2560 hours per year, perform an analysis which characterizes the building's energy consumption.

Solution

The annual energy consumption and energy cost breakdowns are shown in Figure 5.3. It can be seen from the pie charts that although approximately 46% of the energy consumed is natural gas, it only accounts for 16% of the energy costs.

The raw energy consumption and cost figures per m^2 are as follows:

Fuel	Annual energy consumption per m^2	Annual energy cost per m^2
Natural gas	160.965 kWh per m^2	£1.93 per m^2
Electricity	191.163 kWh per m^2	£10.15 per m^2

Using the methodology explained in Section 4.3, the NPI can be established as follows:

Electrical energy used = 955 815 kWh

Gas used = 804 823 kWh

Total energy consumed = 955 815 + 804 823
= 1760 638 kWh

Table 5.2 Energy breakdown figures and normalized performance indicators for various building types in the UK [3]

Building type	Fair NPI (kWh/m²)	Space heating (%)	Domestic hot water (%)	Lighting (%)	Ventilation (%)	Air conditioning (%)	Other (specified) (%)	Other (unspecified) (%)
School with indoor swimming pool	190–240	43	20	9	na	na	22 (Pool)	6
School without indoor swimming pool	250–310	56	25	12	na	na	na	7
Restaurant	410–430	25	15	15	5	na	40 (Catering)	na
Public house	340–470	38	18	12	na	na	11 (Catering)	16
Motorway service area	880–1200	22	32	9	7	na	30 (Catering)	na
Fast food outlet	1450–1750	4	24	1	na	na	70 (Cooking)	1
Supermarket with bakery	1130–1350	23 (includes ventilation)	2	11	na	na	50 (Refrigeration)	3
Office building (naturally ventilated)	230–290	60	8	20	na	na	12 (Electrical consumption)	na
Office building (air conditioned)	250–410	48	6	16	na	29	1 (Office machines)	na
Sports centre (without swimming pool)	200–340	75	3	11	na	na	11 (Fans, pumps, etc.)	na
Swimming pool	1050–1390	10	3	11	45	na	33 (Pool water heating)	na
Church building	88–169	88	na	6	na	na	na	6

Libraries, museums and art galleries	200–280	60	na	18	na	11	na	11
Large hotel	290–420	50	11	9	na	na	18 (Catering)	12
Bank (non-air conditioned)	180–240	67	4	19	na	na	na	10
Cinema	650–780	77	3	2	15	na	3 (Projection equipment)	na
Bingo hall	631–770	65	5	13	5	na	7 (Catering)	5
Prison	550–689	45	25	10	na	na	10 (Catering)	10
Transport depots	311–381	80	4	6	na	na	2 (Catering)	8
Law court	219–300	84	5	8	na	na	1 (Catering)	2
Factory (excluding process energy)	261–369	72	3	15	na	na	na	10
Cold store	500–675	8	na	10	na	na	82 (Refrigeration)	na
Warehouse	150–269	80	2	8	na	na	na	10

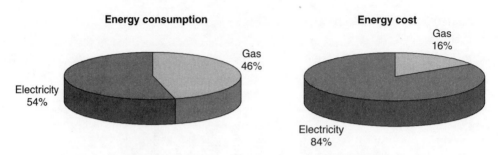

Figure 5.3　Energy consumption and cost breakdowns

From Table 4.5 (see pages 82–3) it can be seen that 75% of the gas consumed can be attributed directly to space heating. Therefore:

$$\text{Space heating energy consumption} = 1760\,638 \times 0.75$$
$$= 1320\,478.5\ \text{kWh}$$

By applying weather and exposure coefficients:

$$\text{Corrected space heating energy consumption} = \frac{1320478.5 \times 2462 \times 1.0}{1932}$$
$$= 1682\,721.6\ \text{kWh}$$
$$\text{Non-heating energy consumption kWh} = 1760\,638 - 1320\,478.5$$
$$= 440\,159.5\ \text{kWh}$$

Therefore

$$\text{Corrected total energy consumption} = 440\,159.5 + 1682\,721.6$$
$$= 2122\,881.1\ \text{kWh}$$

and correcting for occupancy (using data from Table 4.6):

$$\text{Normalized annual energy consumption} = 2122\,881.1 \times \frac{2600}{2560} = 2156\,051.1\ \text{kWh}$$

Therefore

$$\text{NPI} = \frac{2156\,051.1}{5000} = 431.21\ \text{kWh/m}^2$$

Because the calculated NPI of 431.21 kWh/m² is above the upper limit of the 'fair' range (i.e. 410 kWh/m² in Table 5.2), for an air conditioned office building, it can be assumed that the energy performance of the office building is poor.

Using the methodology described in Section 4.5, it is possible to perform a linear regression analysis of the gas data. When such an analysis is performed it yields the following performance equation, which may then be used as the *standard performance* equation for a future monitoring and targeting programme (see Chapter 7).

$$\text{Monthly gas consumption (kWh)} = 31\,521.75 + (220.788 \times \text{Degree days})$$

This equation shows that the monthly gas base-load consumption is 31 521.75 kWh.

5.5 Comprehensive energy audits

It is clear from the analysis presented in Example 5.1 that a considerable amount of useful information can be obtained from a preliminary energy audit. However, without further investigation it is not possible to determine where in the office building the energy is being consumed. In order to do this a more comprehensive audit is required. The analysis techniques used for comprehensive energy audits are essentially much the same as those used for preliminary audits, but the level of detail is much greater. Comprehensive audits require detailed energy surveys to be undertaken, and they often require the installation of additional sub-metering in order to determine accurately component energy flows. Example 5.2 illustrates the auditing benefits which can be gained from installing comprehensive sub-metering.

Example 5.2

Through the installation of sub-meters in the office building in Example 5.1, it has been possible to establish the following data:

Month	Space heating (kWh)	Domestic hot water (kWh)	Catering (kWh)	Lifts (kWh)	Lighting (kWh)	Air conditioning (kWh)	Other (kWh)
January	54 075	13 239	22 696	620	44 016	21 231	2 348
February	61 856	12 924	22 380	610	41 082	16 566	2 054
March	52 888	11 978	22 191	627	33 746	23 659	1 614
April	38 537	11 348	21 435	631	24 942	27 734	11 738
May	12 841	11 663	22 696	636	19 074	68 013	1 511
June	2395	12 293	23 957	640	14 672	88 713	1 907
July	427	11 663	21 750	625	15 112	1011 681	1 819
August	672	10 717	23 011	621	19 074	81 909	1 643
September	14 420	94 57	20 174	615	26 410	59 655	1 555
October	42 507	11 348	22 065	630	32 719	30 134	1 541
November	43 591	11 978	23 011	632	42 549	16 611	1 775
December	67 868	13 554	25 217	628	45 483	21 929	2 083
Totals	392 077	142 163	270 583	7515	358 877	557 835	31 589

From these data it is possible to produce a comprehensive breakdown of annual energy consumption in each of the component areas (see Figure 5.4).

Example 5.2 makes the bold assumption that extensive sub-metering has been installed in the office building for auditing purposes. While this is possible it would be an expensive option. It is therefore often better to employ other techniques to estimate individual energy flows. In some situations it is possible to deduce energy flows by deducting sub-meter readings from main meter readings. However, other situations may require the members of the auditing team to use their skill and judgement to estimate energy flows.

5.5.1 Portable and temporary sub-metering

It may be evident from the outset of an auditing project that the installation of additional sub-metering will yield much useful information about the energy flows within a facility. For example, by placing sub-meters on the energy input side and heat meters on the output side, it is possible to determine the efficiency of individual plant items. Sub-metering should also highlight any imbalances between the consumption recorded by main meters and the total recorded by the sub-meters. If additional sub-meters are deemed necessary, then the

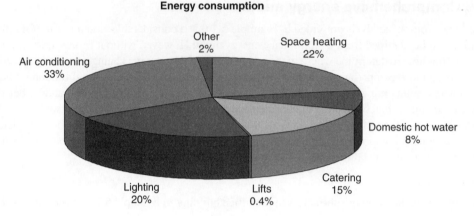

Figure 5.4   Energy consumption breakdown for the office building

additional cost can usually be justified for items of plant or areas with high loads, particularly in situations where little information exists on current energy consumption.

   The process of assessing a facility for additional sub-metering can also highlight shortcomings in the facility's existing metering provision. For example, several buildings may be served
off a single electricity meter. In such a situation, it may be worth considering the installation of
permanent sub-meters, since these would support any future energy management programme.
Permanent metering should also be considered when it is less expensive than the cost of hiring,
installing and removing temporary metering equipment. The installation of permanent or
temporary meters is invasive and involves the shutting off of energy supplies, which in many
circumstances is unsatisfactory. Instead, it is worth considering the use of portable non-invasive
metering. Electricity demand profiles can be monitored using portable clamp-on meters and heat
flows in water pipes can be determined by using clamp-on electronic thermometers and portable
ultrasonic flow meters.

5.5.2 Estimating energy use

In many situations it is either impractical or prohibitively expensive to install comprehensive
sub-metering and so it is necessary to estimate the energy consumption of various items of plant
and equipment. Accurate estimation of equipment energy consumption can be a difficult process
and one which relies on skill and judgement. Nevertheless, it is relatively easy to establish an
upper limit for energy consumption by using eqn (5.1).

$$\text{Annual energy consumption (kWh)} = \frac{Q_{\text{out}}}{\eta} \times T_{\text{h}} \tag{5.1}$$

where Q_{out} is the plant power output (kW), η is the efficiency of plant, and, T_{h} is the number of
operating hours per year.

   While eqn (5.1) may give an upper limit on plant energy consumption it does not provide the
actual operating energy consumption. This problem can be overcome by monitoring the actual
plant energy consumption for a short period of time, using meters, and then multiplying the average measured load by the annual operating time. In the case of electrical equipment this can be
a relatively simple process since the current can be measured using a portable clamp-on meter.

Table 5.3   Seasonal efficiencies for water heating systems [3]

System type	Seasonal efficiency
Gas heater with storage cylinder	52
Gas or oil fired boiler with storage cylinder	56
Hot water cylinder with immersion heater using off-peak electricity	80
Instantaneous gas multi-point heater	62
District heating with local calorifiers	60
District heating with central calorifiers and distribution	56

Space heating energy consumption can be estimated using the heat loss and degree day methods described in Chapter 8. The energy consumed in producing domestic hot water can be determined, if the cold feed water is metered, by using eqn (5.2).

$$\text{Annual energy consumption (kWh)} = \frac{m_{cf} \times c_p \times (t_s - t_{cf})}{\eta_s \times 3600} \tag{5.2}$$

where m_{cf} is the mass of cold feed water used annually (kg), c_p is the specific heat capacity of water (i.e. 4.19 kJ/kgK), t_s is the hot water storage temperature (e.g. 60 °C), t_{cf} is the cold feed water temperature (e.g. 10 °C), and η_s is the seasonal efficiency of system.

The seasonal efficiencies for various domestic hot water producing systems are given in Table 5.3.

If the cold feed water consumption is not known then it is possible to estimate the energy consumed in producing hot water by using eqn (5.3) [4].

$$\text{Annual energy consumption (kWh)} = 0.024 \times q_{hws} \times A_f \times N_w \tag{5.3}$$

Where q_{hws} is the mean power requirement (W/m^2), A_f is the floor area (m^2), and N_w is the number of working days.

The power requirement (q_{hws}) can be determined using Table 5.4. The energy consumed by lighting can be estimated simply by counting the number of luminaire fittings and noting their respective power ratings. With all but tungsten filament lamps, allowances should be made for control gear losses (see Chapter 12). In most buildings lighting energy consumption lies in the region 10–20 W/m^2 [3].

Table 5.4   Mean power requirements for domestic hot water [3]

Building type	Mean power requirement (q_{hws}) (W/m^2)
Office (5 day occupancy)	2.0
Office (6 day occupancy)	2.0
Shop (5 day occupancy)	0.5
Shop (6 day occupancy)	1.0
Factory (5 day occupancy: single shift)	9.0
Factory (6 day occupancy: single shift)	11.0
Factory (7 day occupancy: multiple shift)	12.0
Warehouse	1.0
Residential buildings	17.5
Hotels	8.0
Hospitals	29.0
Education	2.0

5.6 Energy surveys

Energy surveys are an integral part of the auditing process. They enable the auditors to understand the energy flows within facilities and to identify energy wastage. Surveys can be comprehensive, looking in depth at all aspects of a facility's energy consumption, or targeted, in which case they only cover certain specific issues. The main objectives of any energy survey should be to:

- Determine the energy performance of a facility, or in the case of a targeted survey, targeted items of plant and equipment.
- Identify and quantify the principal energy flows.
- Identify and quantify achievable energy cost savings.
- Produce costed recommendations to achieve energy cost savings.
- Make recommendations on the future energy management of the facility.

Energy surveys, with the exception of specifically targeted surveys, should cover all aspects relating to a facility's or organization's energy consumption. This will involve detailed surveys of:

- The management and operation characteristics of a facility or organization.
- The energy supply to an organization's various facilities.
- The energy use within a facility.
- The plant and equipment within a facility.
- The fabric of the organization's buildings.

5.6.1 Management and operating characteristics

The management culture within an organization can have a great influence on energy consumption. It is therefore important to determine the management structure and practices relating to energy procurement and consumption. In particular, it is important to identify cost centres clearly, where the managers accountable for operating costs can be made individually responsible for energy consumption. Maintenance practices can also have a direct influence on energy consumption, so it is important to establish the frequency and quality of the maintenance procedures, and to identify new maintenance measures which could improve the energy performance of plant and equipment.

At the auditing stage it is important to survey the operating practices within an organization or facility. Detailed data should be gathered on such factors as:

- The use of a particular space or building.
- The mechanical and electrical services within the building.
- The number and type of occupants. Particular attention should be paid to any special characteristics of occupants. For example, in rooms containing smokers, windows are often opened with the result that space heating costs are increased.
- The occupancy patterns of building or space.
- The environmental conditions within a space or building. This will include air temperature, dry resultant temperature, relative humidity and illuminance levels.
- The operating practices of major items of plant and equipment.

Knowledge of the above issues will give the auditor a good understanding of how and why energy is consumed within a facility. Particular attention should be paid to situations where actual practice deviates from that stated by the management of an organization. For example, if rooms which are supposed to be heated to 21 °C are in fact heated to 23 °C, then energy is being wasted on over-heating spaces.

5.6.2 Energy supply

It is important to identify the tariffs and supply contracts under which any organization purchases its energy. This will enable the energy auditing team to establish whether or not a particular organization is purchasing energy at a low price. If an organization is paying a higher than necessary price for its energy, then the auditor should recommend a change of tariff or fuel supplier.

Because electricity tariffs usually include some form of demand charge, it is important that an organization selects the correct electricity tariff to suite its load profile. Therefore, the audit process should include a survey of the electrical load profile of a facility. For relatively minor loads it may be sufficient to take meter readings at the beginning and end of a selected period, with intermediate readings taken during the daytime, nighttime and at the weekend. This will give a good indication of when electrical energy is being consumed and should assist the auditor in recommending an appropriate tariff. For larger electrical loads, it is important to survey the load profile accurately. This can be achieved by using a portable meter to determine demand and consumption at 30-minute intervals over a selected period. Any large peaks in load should be identified and further investigations made in order to establish their cause.

With electricity supply, it is important to determine the power factor of a facility. Many items of equipment, such as fluorescent lamps and electric motors, produce a poor power factor (i.e. a decoupling of the current and the voltage so that they become out of phase with each other). This results in higher than expected electricity bills. If poor power factors are found in a facility then it may be worth considering the installation of power factor correction equipment (see Chapter 12).

5.6.3 Plant and equipment

Major items of plant, such as boilers and refrigeration chillers, convert energy from one form to another. In doing so energy is wasted. For example, in boilers much of the heat produced by the combustion process can be wasted by allowing it to escape with the flue gases. The more efficient an item of plant, the less energy is wasted. Major items of plant and equipment should therefore be surveyed in order to determine their operating efficiency. It is also important to survey their respective pipe distribution networks since these too can be a major source of energy wastage.

The subjects of energy efficient heating and refrigeration are dealt with in some detail in Chapters 8 and 11, so the issues involved in a plant survey are only briefly summarized here. With boilers it is essential that they be 'tuned' so that flue gas heat losses are minimized. This involves sampling the CO_2 or O_2 content of the flue gases, and adjusting the burner settings so that excess O_2 is minimized, whilst still ensuring that complete combustion takes place. In addition, it is important to identify whether or not flue gas heat recovery is feasible.

The efficiency of refrigeration plant is measured by its coefficient of performance (COP). The higher the COP, the greater the efficiency of the machine. COP varies with the cooling load and the external air conditions. It is therefore necessary to meter the energy input and output over a period of time, if the average COP is to be identified. Note should be taken of the operating pattern of refrigeration plant and also of how it is controlled, since this will tell the auditor much about the operation of the facility. If both boilers and refrigeration chillers are operating at the same time, it could be that the two systems are fighting each other. If this is the case, then the control logic and settings of the system will need adjustment. The feasibility of recovering heat from the hot condenser gases should also be investigated.

In many facilities much energy is wasted from hot water, chilled water and steam distribution pipework because of inadequate or poor quality insulation. Pipework systems should therefore be inspected to establish the quality of the insulation and also to identify any leaks.

Plant surveys should allow for the fact that mechanical equipment has a finite working life and that efficiency often deteriorates badly when plant is old. Therefore one of the important outcomes of such a survey should be a recommendation for the planned replacement of older plant. In many situations it is much more cost effective to replace old plant, rather than renovate it.

5.6.4 Building fabric

Although the subject of heat loss through building fabric is covered in detail in Chapter 8, a few words on the subject are timely here. It is important to note the age, size, shape and orientation of the buildings within a facility, since these are all factors which affect energy consumption. In particular, areas of greatest heat loss should be identified. Greatest heat loss will always occur where the building fabric has a high U value (see Chapter 8). If 'as built' drawings exist, these can be used to determine elemental U values. The use of portable infrared thermography can also be very useful, since this enables areas of high heat loss to be instantly identified.

When surveying buildings, it should be appreciated that large amounts of heat can be lost by excess ventilation (see Chapter 8). This is particularly the case in older properties with old fenestration systems. Particular attention should therefore be paid to any poorly fitting window and door frames, or to any space where windows and outside doors remain open for any length of time.

5.7 Recommendations

The energy auditing process should enable recommendations to be made, which will result in cost savings. Although the precise nature of these recommendations will depend on the particular application in question, they can broadly be classified as follows:

(i) *Reducing energy costs by tariff negotiation*: Electricity and gas are supplied either through published tariffs or through negotiated supply contracts. Not all tariffs and supply contracts are suited to every organization and some are better than others. It may therefore be possible to reduce energy costs simply by changing tariff or negotiating a more beneficial supply contract.

(ii) *Good maintenance and work practices*: Energy can often be saved at no capital expense simply by 'good housekeeping' (i.e. improving maintenance procedures and implementing good work practices).

(iii) *Retrofitting and tuning systems*: Energy is often wasted as systems age, because components wear out or become damaged. Also, the controls associated with these systems are often inappropriate or poorly set-up with the result that systems perform inefficiently. Significant energy savings can be achieved through modest capital investment to retrofit and re-tune inefficient installations.

(iv) *Capital investment*: In many situations the poor condition of plant and infrastructure makes refurbishment a futile exercise. Under these circumstances major capital investment is required to replace existing plant. In this case, it is often worthwhile reappraising the situation to determine whether or not an alternative installation might be more appropriate. An existing boiler installation could be replaced by a combined heat and power (CHP) plant, thus reducing the need to buy in electrical power. Such measures usually involve large capital investment and careful financial appraisal is therefore required.

5.8 The audit report

The audit process should identify potential energy management opportunities. Since exploitation of these opportunities often involves capital expenditure, maximum effort should be put into investigating those measures which will yield the greatest cost savings. Those energy management opportunities which result in lesser savings should be given a low priority. However, all energy management opportunities, identified through the auditing process should be clearly stated in the final audit report, together with cost benefit calculations to justify them. The final audit report should include:

- A description of the facility, including layout drawings, construction details, hours of operation, equipment lists and any relevant materials and product flows.
- A description of the various utility tariffs or contracts used.
- A presentation of all the energy data gathered, together with any relevant analysis.
- A detailed statement of potential energy management opportunities, together with supporting cost-benefit analysis calculations.
- An energy management action plan for the future operation of the facility. This may include an implementation schedule for the recommended energy management opportunities and a programme for the ongoing energy monitoring and targeting of the facility.

Although the audit report should contain detailed technical information, it is important to remember that its primary purpose is to communicate the principal findings of the audit to an organization's senior management, many of whom may have little understanding of energy matters. It is therefore advisable to include a short executive summary, giving a brief synopsis of the report and highlighting its major findings and recommendations.

References

1. How to hire an energy auditor to identify energy efficiency projects. (2000). California Energy Commission. January.
2. Aspects of energy management. (1995). General Information Report 12, Department of the Environment. May.
3. Energy audits and surveys. (1991). CIBSE Applications Manual AM5.
4. Measurement of energy consumption: CIBSE Building Energy Code Part 4. (1982). CIBSE.

Bibliography

Capehurst, B. L., Turner, W. C. and Kennedy, W. J. (1997). *Guide to energy management* (Chapter 2). The Fairmont Press, Prentice Hall.
Eastop, T. D. and Croft, D. R. (1990). *Energy efficiency for engineers and technologists* (Chapter 9). Longman Scientific & Technical.
Energy audits and surveys (1991). CIBSE Applications Manual AM5.
Energy audits for buildings, Fuel Efficiency Booklet 1 (1993). Department of the Environment.
How to hire an energy auditor to identify energy efficiency projects (2000). California Energy Commission. January.
Moss, K. J. (1997). *Energy management and operating costs in buildings* (Chapter 9). E & FN Spon.
Thumann, A. and Mehta, P. D. (1997). *Handbook of energy engineering* (Chapter 3). The Fairmont Press, Prentice Hall.

6

Project investment appraisal

It is important to justify any capital investment project by carrying out a financial appraisal. The financial issues associated with capital investment in energy saving projects are investigated in this chapter. In particular, the discounted cash flow techniques of *net present value* and *internal rate of return* are presented and discussed.

6.1 Introduction

When planning an energy efficiency or energy management project, the costs involved should always be considered. Therefore, as with any other type of investment, energy management proposals should show the likely return on any capital that is invested. Consider the case of an energy consultant who advises the senior management of an organization that capital should be invested in new boiler plant. Inevitably, the management of the organization would enquire:

- How much will the proposal cost?
- How much money will be saved by the proposal?

These are, of course, not unreasonable questions, since within any organization there are many 'worthy causes', each of which requires funding and it is the job of senior management to invest capital where it is going to obtain the greatest return. In order to make a decision about any course of action, management needs to be able to appraise all the costs involved in a project and determine the potential returns. This however, is not quite as straightforward as it might first appear. The capital value of plant or equipment usually decreases with time and it often requires more maintenance as it gets older. If money is borrowed from a bank to finance a project, then interest will have to be paid on the loan. Inflation too will influence the value of any future energy savings that might be achieved. It is therefore important that the cost appraisal process allows for all these factors, with the aim of determining which investments should be undertaken, and of optimizing the benefits achieved. To this end a number of accounting and financial appraisal techniques have been developed which help managers make correct and objective decisions. It is these financial appraisal techniques which are introduced and discussed in this chapter.

6.2 Fixed and variable costs

When appraising the potential costs involved in a project it is important to understand the difference between fixed and variable costs. Variable costs are those which vary directly with the output of a particular plant or production process, such as fuel costs. Fixed costs are those costs, which are not dependent on plant or process output, such as site-rent and insurance. The total cost of any project is therefore the sum of the fixed and variable costs. Example 6.1 illustrates how both fixed and variable costs combine to make the total operating cost.

Example 6.1

Determine the total cost of a diesel generator operating over a 5-year period. Assume that the capital cost of the generator is £15 000, the annual output is 219 MWh, and the maintenance costs are £500 per annum. The cost of producing each unit of electricity is 3.5 p/kWh.

Solution

Item	Type of cost	Calculation	Cost (£)
Capital cost of generator	Fixed	n.a.	15 000.00
Annual maintenance	Fixed	£ 500 × 5	2 500.00
Fuel cost	Variable	219 000 × 0.035	7 665.00
		Total cost =	25 165.00

From Example 6.1 it can be seen that the fixed costs represent 69.5% of the total cost. In fact, the annual electricity output of 219 MWh assumes that the plant is operating with an average output of 50 kW. If this output were increased to an average of 70 kW, then the fuel cost would become £10 731, with the result that the fixed costs would drop to 62% of the total. Clearly, the average unit cost of production decreases as output increases.

The concept of fixed and variable costs can be used to determine the break-even point for a proposed project. The break-even point can be determined by using eqn (6.1).

$$UC_{util} \times W_{av} \times n = FC + (UC_{prod} \times W_{av} \times n) \tag{6.1}$$

where UC_{util} is the unit cost per kWh of bought in energy (£/kWh), UC_{prod} is the unit cost per kWh of produced energy (£/kWh), FC is the fixed costs (£), W_{av} is the average power output (or consumption) (kW), and n is the number of hours of operation (h).

Example 6.2

Assuming that electricity bought from a local utility company costs an average of 6.1 p/kWh, determine the break-even point for the generator described in Example 6.1, when:

(i) the average output is 50 kW; and
(ii) the average output is 70 kW.

Solution

(i) Assuming that the average output of the generator is 50 kW.

$$0.061 \times 50 \times n = (15\ 000 + 2500) + (0.035 \times 50 \times n)$$
$$\therefore \quad n = 13\ 461.5 \text{ hours}$$

(ii) Assuming that the average output of the generator is 70 kW.

$$0.061 \times 70 \times n = (15\ 000 + 2500) + (0.035 \times 70 \times n)$$
$$\therefore \quad n = 9615.4 \text{ hours}$$

Clearly, increasing the average output of the generator significantly reduces the break-even time for the project. This is because the capital investment (i.e. the generator) is being better utilized.

6.3 Interest charges

In order to finance projects, organizations often borrow money from banks or other lending organizations. Projects financed in this way cost more than similar projects financed from an organization's own funds, because interest charges must be paid on the loan. It is therefore important to understand how interest charges are calculated. Interest charges can be calculated by lending organizations in two different ways: simple interest and compound interest.

(i) *Simple interest*: If simple interest is applied, then charges are calculated as a fixed percentage of the capital that is borrowed. A fixed interest percentage is applied to each year of the loan and repayments are calculated using eqn (6.2).

$$TRV = LV + \left(\frac{IR}{100} \times LV \times P\right) \qquad (6.2)$$

where TRV is the total repayment value (£), LV is the value of initial loan (£), IR is the interest rate (%), and P is the repayment period (years).

(ii) *Compound interest*: Compound interest is usually calculated annually (although this is not necessarily the case). The interest charged is calculated as a percentage of the outstanding loan at the end of each time period. It is termed 'compound' because the outstanding loan is the sum of the unpaid capital and the interest charges up to that point. The value of the total repayment can be calculated using eqn (6.3).

$$TRV = LV \times \left(1 + \frac{IR}{100}\right)^P \qquad (6.3)$$

The techniques involved in calculating simple and compound interest are illustrated in Example 6.3.

Example 6.3

A company borrows £ 50 000 to finance a new boiler installation. If the interest rate is 9.5% per annum and the repayment period is 5 years, determine the value of the total repayment and the monthly repayment value, assuming:

(i) that simple interest is applied; and
(ii) that compound interest is applied.

Solution

(i) *Assuming simple interest*:

$$\text{Total repayment} = 50\ 000 + \left(\frac{9.5}{100} \times 50\ 000 \times 5\right) = £73\ 750.00$$

$$\text{Monthly repayment} = \frac{73\ 750.00}{(5 \times 12)} = £1229.17$$

(ii) *Assuming compound interest*:

$$\text{Repayment at end of year 1} = 50\ 000 + \left(\frac{9.5}{100} \times 50\ 000\right) = £54\ 750.00$$

and

$$\text{Repayment at end of year 2} = 54\,750 + \left(\frac{9.5}{100} \times 54\,750\right) = £59\,951.25$$

Similarly, the repayments at the end of years 3, 4 and 5 can be calculated:

Repayment at end of year 3 = £ 65 646.62
Repayment at end of year 4 = £ 71 883.05
Repayment at end of year 5 = £ 78 711.94

Alternatively, eqn (6.3) can be used to determine the compound interest repayment value.

$$\text{Total repayment value} = 50\,000 \times \left(1 + \frac{9.5}{100}\right)^5 = £78\,711.94$$

$$\text{Monthly repayment} = \frac{78\,711.94}{(5 \times 12)} = £1311.87$$

It can be seen that by using compound interest, the lender recoups an additional £4962. Not surprisingly lenders usually charge compound interest on loans.

6.4 Payback period

Probably the simplest technique which can be used to appraise a proposal is payback analysis. The payback period can be defined as 'the length of time required for the running total of net savings before depreciation to equal the capital cost of the project' [1]. In theory, once the payback period has ended, all the project capital costs will have been recouped and any additional cost savings achieved can be seen as clear 'profit'. Obviously, the shorter the payback period, the more attractive the project becomes. The length of the maximum permissible payback period generally varies with the business culture concerned. In some countries, payback periods in excess of 5 years are considered acceptable, whereas in other countries, such as the in the UK, organizations generally impose payback periods of less than 3 years.
The payback period can be calculated using eqn (6.4).

$$PB = \frac{CC}{AS} \tag{6.4}$$

where PB is the payback period (years), CC is the capital cost of the project (£), and AS is the annual net cost saving achieved (£).

The annual net cost saving (AS) is the cost saving achieved after all the operational costs have been met.

Example 6.4

A new combined heat and power (CHP) installation is expected to reduce a company's annual energy bill by £8100. If the capital cost of the new boiler installation is £37 000, and the annual maintenance and operating costs are £700, what will be the expected payback period for the project?

Solution

$$PB = \frac{37\,000}{8100 - 700} = 5.0 \text{ years}$$

6.5 Discounted cash flow methods

The payback method is a simple technique which can easily be used to provide a quick evaluation of a proposal. However, it has a number of major weaknesses:

- The payback method does not consider savings that are accrued after the payback period has finished.
- The payback method does not consider the fact that money, which is invested, should accrue interest as time passes. In simple terms there is a 'time value' component to cash flows. Thus a £100 today is more valuable than £100 in 10 years' time.

In order to overcome these weaknesses a number of discounted cash flow techniques have been developed, which are based on the fact that money invested in a bank will accrue annual interest. The two most commonly used techniques are the 'net present value' and the 'internal rate of return' methods.

6.5.1 Net present value method

The net present value method considers the fact that a cash saving (often referred to as a 'cash flow') of £1000 in year 10 of a project will be worth less than a cash flow of £1000 in year 2. The net present value method achieves this by quantifying the impact of time on any particular future cash flow. This is done by equating each future cash flow to its current value today, in other words determining the *present value* of any future cash flow. The *present value* (PV) is determined by using an assumed interest rate, usually referred to as a *discount rate*. Discounting is the opposite process to compounding. Compounding determines the future value of present cash flows, whereas discounting determines the *present value* of future cash flows.

In order to understand the concept of *present value*, consider the case described in Example 6.4. If instead of installing a new CHP system, the company invested £37 000 in a bank at an annual interest rate of 8%, then:

The value of the sum at the end of year 1 = 37 000 + (0.08 × 37 000) = £39 960.00

and

The value of the sum at the end of year 2 = 39 960 + (0.08 × 39 960) = £43 156.80

The value of the investment would grow as compound interest is added, until after n years the value of the sum would be:

$$FV = D \times \left(1 + \frac{IR}{100}\right)^n \tag{6.5}$$

where FV is the future value of investment (£), and D is the value of initial deposit (or investment) (£).

So after 5 years the future value of the investment would be:

$$FV = 37\,000 \times \left(1 + \frac{8}{100}\right)^5 = £54\,365.14$$

So in 5 years the initial investment of £37 000 will accrue £17 365.14 in interest and will be worth £54 365.14. Alternatively, it could equally be said that £54 365.14 in 5 years time is worth £37 000 now (assuming an annual interest rate of 8%). In other words the *present value* of £54 365.14 in 5 years time is £37 000 now. The *present value* of an amount of money at any specified time in the future can be determined by eqn (6.6).

$$PV = S \times \left(1 + \frac{IR}{100}\right)^{-n}$$

where PV is the present value of S in n years time (£), and S is the value of cash flow in n years time (£).

The *net present value* method calculates the *present value* of all the yearly cash flows (i.e. capital costs and net savings) incurred or accrued throughout the life of a project, and summates them. Costs are represented as a negative value and savings as a positive value. The sum of all the present values is known as the *net present value* (NPV). The higher the *net present value*, the more attractive the proposed project.

The *present value* of a future cash flow can be determined using eqn (6.6). However, it is common practice to use a *discount factor* (DF) when calculating present value. The discount factor is based on an assumed discount rate (i.e. interest rate) and can be determined by using eqn (6.7).

$$DF = \left(1 + \frac{IR}{100}\right)^{-n} \tag{6.7}$$

The product of a particular cash flow and the *discount factor* is the *present value*.

$$PV = S \times DF \tag{6.8}$$

The value of various *discount factors* computed for a range of discount rates (i.e. interest rates) is shown in Table 6.1. Example 6.5 illustrates the process involved in a *net present value* analysis.

Table 6.1 Computed discount factors

Year	Discount rate % (or interest rate %)							
	2	4	6	8	10	12	14	16
0	1.000	1.000	1.000	1.000	1.000	1.000	1.000	1.000
1	0.980	0.962	0.943	0.926	0.909	0.893	0.877	0.862
2	0.961	0.925	0.890	0.857	0.826	0.797	0.769	0.743
3	0.942	0.889	0.840	0.794	0.751	0.712	0.675	0.641
4	0.924	0.855	0.792	0.735	0.683	0.636	0.592	0.552
5	0.906	0.822	0.747	0.681	0.621	0.567	0.519	0.476
6	0.888	0.790	0.705	0.630	0.564	0.507	0.456	0.410
7	0.871	0.760	0.665	0.583	0.513	0.452	0.400	0.354
8	0.853	0.731	0.627	0.540	0.467	0.404	0.351	0.305
9	0.837	0.703	0.592	0.500	0.424	0.361	0.308	0.263
10	0.820	0.676	0.558	0.463	0.386	0.322	0.270	0.227
11	0.804	0.650	0.527	0.429	0.350	0.287	0.237	0.195
12	0.788	0.625	0.497	0.397	0.319	0.257	0.208	0.168
13	0.773	0.601	0.469	0.368	0.290	0.229	0.182	0.145
14	0.758	0.577	0.442	0.340	0.263	0.205	0.160	0.125
15	0.743	0.555	0.417	0.315	0.239	0.183	0.140	0.108
16	0.728	0.534	0.394	0.292	0.218	0.163	0.123	0.093
17	0.714	0.513	0.371	0.270	0.198	0.146	0.108	0.080
18	0.700	0.494	0.350	0.250	0.180	0.130	0.095	0.069
19	0.686	0.475	0.331	0.232	0.164	0.116	0.083	0.060
20	0.673	0.456	0.312	0.215	0.149	0.104	0.073	0.051

Example 6.5

Using the *net present value* analysis technique, evaluate the financial merits of the two proposed projects shown in the table below. Assume an annual discount rate of 8% for each project.

	Project 1	Project 2
Capital cost (£)	30 000.00	30 000.00
Year	**Net annual saving (£)**	**Net annual saving (£)**
1	+6 000.00	+6 600.00
2	+6 000.00	+6 600.00
3	+6 000.00	+6 300.00
4	+6 000.00	+6 300.00
5	+6 000.00	+6 000.00
6	+6 000.00	+6 000.00
7	+6 000.00	+5 700.00
8	+6 000.00	+5 700.00
9	+6 000.00	+5 400.00
10	+6 000.00	+5 400.00
Total net saving at end of year 10	**+60 000.00**	**+60 000.00**

Solution

The annual cash flows should be multiplied by the annual *discount factors* for a rate of 8% to determine the annual *present values*, as shown in the table below.

Year	Discount factor for 8% (a)	Project 1 Net savings (£) (b)	Project 1 Present value (£) (a × b)	Project 2 Net savings (£) (c)	Project 2 Present value (£) (a × c)
0	1.000	−30 000.00	−30 000.00	−30 000.00	−30 000.00
1	0.926	+6 000.00	+5 556.00	+6 600.00	+6 111.60
2	0.857	+6 000.00	+5 142.00	+6 600.00	+5 656.20
3	0.794	+6 000.00	+4 764.00	+6 300.00	+5 002.20
4	0.735	+6 000.00	+4 410.00	+6 300.00	+4 630.50
5	0.681	+6 000.00	+4 086.00	+6 000.00	+4 086.00
6	0.630	+6 000.00	+3 780.00	+6 000.00	+3 780.00
7	0.583	+6 000.00	+3 498.00	+5 700.00	+3 323.10
8	0.540	+6 000.00	+3 240.00	+5 700.00	+3 078.00
9	0.500	+6 000.00	+3 000.00	+5 400.00	+2 700.00
10	0.463	+6 000.00	+2 778.00	+5 400.00	+2 500.20
			NPV = +10 254.00		NPV = +10 867.80

It can be seen that over a 10-year life-span the *net present value* for Project 1 is £10 254.00, while for Project 2 it is £10 867.80. Therefore Project 2 is the preferential proposal.

The whole credibility of the *net present value* method depends on a realistic prediction of future interest rates, which can often be unpredictable. It is prudent therefore to set the discount rate slightly above the interest rate at which the capital for the project is borrowed. This will ensure that the overall analysis is slightly pessimistic, thus acting against the inherent uncertainties in predicting future savings.

6.5.2 Internal rate of return method

It can be seen from Example 6.5 that both projects returned a positive net present value over 10 years, at a discount rate of 8%. However, if the discount rate were reduced there would come a point when the *net present value* would become zero. It is clear that the discount rate which must be applied, in order to achieve a net present value of zero, will be higher for Project 2 than for Project 1. This means that the average rate of return for Project 2 is higher than for Project 1, with the result that Project 2 is the better proposition. The discount rate which achieves a *net present value* of zero is known as the *internal rate of return* (IRR). The higher the *internal rate of return*, the more attractive the project.

Example 6.6 illustrates how an *internal rate of return* analysis is performed.

Example 6.6

A proposed project requires an initial capital investment of £20 000. The cash flows generated by the project are shown in the table below.

Year	Cash flow (£)
0	−20 000.00
1	+6 000.00
2	+5 500.00
3	+5 000.00
4	+4 500.00
5	+4 000.00
6	+4 000.00

Given the above cash flow data determine the internal rate of return for the project.

Solution

The net present value should be calculated for a range of discount rates, as shown below.

Year	Cash flow (£)	8% discount rate		12% discount rate		16% discount rate	
		Discount factor	Present value (£)	Discount factor	Present value (£)	Discount factor	Present value (£)
0	−20 000	1.000	−20 000	1.000	−20 000	1.000	−20 000
1	6 000	0.926	5 556	0.893	5 358	0.862	5 172
2	5 500	0.857	4 713.5	0.797	4 383.5	0.743	4 086.5
3	5 000	0.794	3 970	0.712	3 560	0.641	3 205
4	4 500	0.735	3 307.5	0.636	2 862	0.552	2 484
5	4 000	0.681	2 724	0.567	2 268	0.476	1 904
6	4 000	0.630	2 520	0.507	2 028	0.410	1 640
			NPV = 2 791		NPV = 459.5		NPV = −1 508.5

It can clearly be seen that the discount rate which results in the net present value being zero lies somewhere between 12% and 16%. The exact internal rate of return can be found by plotting the net present values on a graph, as shown in Figure 6.1.

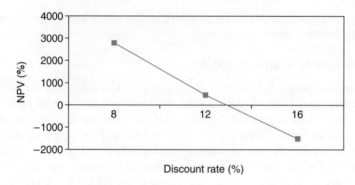

Figure 6.1 NPV versus discount rate

Figure 6.1 shows that the *internal rate of return* for the project is 12.93%. At first sight both the *net present value* and the *internal rate of return* methods look very similar, and in some respects they are. Yet there is an important difference between the two. The *net present value* method is essentially a comparison tool, which enables a number of projects to be compared, while the *internal rate of return* method is designed to assess whether or not a single project will achieve a target rate of return.

6.5.3 Profitability index

Another technique which can be used to evaluate the financial viability of projects is the *profitability index*. The *profitability index* can be defined as:

$$\text{Profitability index} = \frac{\text{Sum of the discounted net savings}}{\text{Capital costs}} \tag{6.9}$$

The higher the *profitability index*, the more attractive the project. The application of the *profitability index* is illustrated in Example 6.7.

Example 6.7

Determine the *profitability index* for the projects outlined in Example 6.5.

Solution

$$\text{Project 1:} \quad \text{Profitability index} = \frac{40\ 254.00}{30\ 000.00} = 1.342$$

$$\text{Project 2:} \quad \text{Profitability index} = \frac{40\ 867.80}{30\ 000.00} = 1.362$$

Project 2 is therefore a better proposal than Project 1.

6.6 Factors affecting analysis

Although the examples 6.5 and 6.6 illustrate the basic principles associated with the financial analysis of projects, they do not allow for the following important considerations:

- The capital value of plant and equipment generally depreciates over time.
- General inflation reduces the value of savings as time progresses. For example, £100 saved in 1 year's time will be worth more than £100 saved in 10 years' time.

The capital depreciation of an item of equipment can be considered in terms of its salvage value at the end of the analysis period. Example 6.8 illustrates this point.

Example 6.8

It is proposed to install a heat recovery device in a factory building. The capital cost of installing the device is £20 000 and after 5 years its salvage value is £1500. If the savings accrued by the heat recovery device are as shown below, determine the net present value after 5 years. Assume a discount rate of 8%.

Data:

Year	1	2	3	4	5
Saving (£)	7000	6000	6000	5000	5000

Solution

Year	Discount factor for 8% (a)	Capital investment (£) (b)	Net savings (£) (c)	Present value (£) (a) $\times$ (b + c)
0	1.000	−20 000.00		−20 000.00
1	0.926		+7000.00	+6 482.00
2	0.857		+6000.00	+5 142.00
3	0.794		+6000.00	+4 764.00
4	0.735		+5000.00	+3 675.00
5	0.681	+1 500.00	+5000.00	+4 426.50
				NPV = +4 489.50

It is evident that over a 5-year life-span the *net present value* of the project is £4489.50. Had the salvage value of the equipment not been considered, the *net present value* of the project would have been only £3468.00.

6.6.1 Real value

Inflation can be defined as the 'rate of increase in the average price of goods and services' [1]. In the UK, inflation is expressed in terms of the *retail price index* (RPI), which is determined centrally and reflects average inflation over a range of commodities. Because of inflation, the *real value* of cash flows decrease with time. The real value of a sum of money (S) realized in n years time can be determined by using eqn (6.10).

$$RV = S \times \left(1 + \frac{R}{100}\right)^{-n} \tag{6.10}$$

where RV is the real value of S realized in n years time (£), S is the value of cash flow in n years time (£), and R is the inflation rate (%).

As with the 'discount factor' it is common practice to use an 'inflation factor' when assessing the impact of inflation on a project. The inflation factor can be determined by using eqn (6.11).

$$IF = \left(1 + \frac{R}{100}\right)^{-n} \tag{6.11}$$

The product of a particular cash flow and the inflation factor is the *real value* of the cash flow.

$$RV = S \times IF \tag{6.12}$$

The application of inflation factors is considered in Example 6.9.

Example 6.9

Recalculate the net present value of the energy recovery scheme in Example 6.8, assuming that the discount rate remains at 8% and that the rate of inflation is 5%.

Solution

Because of inflation;

Real interest rate = Discount rate − Rate of inflation

∴   Real interest rate = 8 − 5 = 3%

Year	Capital investment (£)	Net real savings (£)	Inflation factor for 5%	Net real savings (£)	Real discount factor for 3%	Present value (£)
0	−20 000.00		1.000	−20 000.00	1.000	−20 000.00
1		+7 000.00	0.952	+6 664.00	0.971	+6 470.74
2		+6 000.00	0.907	+5 442.00	0.943	+5 131.81
3		+6 000.00	0.864	+5 184.00	0.915	+4 743.36
4		+5 000.00	0.823	+4 115.00	0.888	+3 654.12
5	+1 500.00	+5 000.00	0.784	+5 096.00	0.863	+4 397.85
					NPV =	+4 397.88

Example 6.9 shows that when inflation is assumed to be 5%, the *net present value* of the project reduces from £4489.50 to £4397.88. This is to be expected, because general inflation will always erode the value of future 'profits' accrued by a project.

References

1. Eastop, T. D. and Croft, D. R. (1990). *Energy efficiency for engineers and technologists* (Chapter 2). Longman Scientific & Technical.

Bibliography

Capehurst, B. L., Turner, W. C. and Kennedy, W. J. (1997). *Guide to energy management* (Chapter 4). The Fairmont Press, Prentice Hall.

Eastop, T. D. and Croft, D. R. (1990). *Energy efficiency for engineers and technologists* (Chapter 2). Longman Scientific & Technical.

Investment appraisal for industrial energy efficiency (1993). Good Practice Guide 69. DOE.

Moss, K. J. (1997). *Energy Management and Operating Costs in Buildings* (Chapter 8). E & FN Spon.

Sizer, J. (1979). *An Insight into Management Accounting* (Chapter 8). Pelican.

Thumann, A. and Mehta, P. D. (1997). *Handbook of energy engineering* (Chapter 2). The Fairmont Press, Prentice Hall.

7

Energy monitoring, targeting and waste avoidance

The concept of monitoring and targeting is discussed in this chapter and techniques for collecting and analysing energy data are explained. A variety of reporting techniques is also presented. The subject of waste avoidance is discussed and practical examples given on how to diagnose and eliminate wasteful energy practices.

7.1 The concept of monitoring and targeting

It is possible to establish the existing energy consumption of a facility or organization through an energy audit (see Chapter 5). However, this only produces a 'picture' of past energy consumption. In order to keep control of subsequent energy consumption, it is necessary to initiate a monitoring programme, although by itself it is only of limited value, as it simply records energy consumption. To achieve improvements in energy performance a targeting programme, in which targets are set, must accompany the monitoring process and planned improvements made. The key elements of a *monitoring and targeting* (M&T) programme are as follows [1]:

- The establishment of Energy Account Centres (EACs) within an organization. These may be departments, processes or cost centres. Operational managers should be accountable for the energy consumption of the EACs for which they are responsible.
- The establishment of *standard energy performance* benchmarks for each EAC. Standard energy performance relates energy consumption to a variable, such as degree days or production output. Standard performance is established through regression analysis of past energy data (see Chapter 4) and it provides a base line for the assessment of future energy performance.
- Monitoring the energy consumption of each EAC within an organization. This involves setting up procedures to ensure the regular collection of reliable energy data.
- The establishment of energy targets for each EAC. Energy cost savings can only be achieved if improvements are made on standard performance. Achievable targets should therefore be set which improve on standard performance.

- Energy management reports should be produced for each EAC on a regular basis. These reports provide the stimulus for improved energy performance, and should also quantify any improvements that are achieved.

It is important that any proposed M&T programme be designed to suit the needs of its host organization. From an energy point of view, organizations can be characterized in a variety of ways. One useful classification method is by the number of sites covered and the level of metering adopted, as follows [1]:

- Single site with central utility metering;
- Single site with sub-metering;
- Multisite with central utility metering;
- Multisite with sub-metering.

Single sites with central utility metering are probably best treated as a single EAC, while the introduction of sub-metering enables such sites to be broken-up into a number of separate EACs. Where organizations have a number of separate properties, each with central utility meters, the sites should be treated as separate EACs. If organizations have multiple properties, each containing sub-metering, then it should be possible to divide each site into a number of separate EACs.

7.2 Computer-based monitoring and targeting

The use of specially designed computer software is advisable when operating an M&T programme. Computers should not be seen as a replacement for the energy manager, but simply as tools which enable large amounts of data to be stored and analysed in a short period of time. A number of energy management software packages are commercially available, with varying degrees of complexity (discussion of which is beyond the scope of this book). They all tend to share the following generic features:

- A database facility, which is capable of storing and organizing large quantities of data collected over a long period of time.
- The ability to record energy data for all utility types, including data taken from both meters and invoices.
- The ability to handle complex utility tariffs. Tariffs vary from country to country, and are becoming increasingly complex as competition is introduced into the utilities sector.
- The ability to handle other variables such as degree days and production data.
- A data analysis facility. This is achieved by incorporating statistical analysis software into the energy management software.
- A reporting facility, which is capable of quickly producing energy management reports.

With the more sophisticated energy management packages it is possible to interface the software with Building Management Systems (BMS), so that energy data can be automatically recorded on a regular basis (e.g. hourly).

One of the great advantages of computer-based systems is their database facility, which enables historical data and data from many sources to be instantly compared. This facility is particularly useful when comparing site energy costs on a utility basis and enables energy managers quickly to assess the relative performance of various EACs. In this way EACs which are under-performing can be quickly identified and remedial action taken.

7.3 Monitoring and data collection

An integral element of any M&T programme is data collection. Ultimately, the accuracy of an energy audit or any M&T programme depends on the collection and input of good quality data. In order to ensure accuracy, robust data collection procedures must be established, but problems can arise for a variety of reasons. If too few data are used, any analysis will be meaningless. Most systems therefore require at least 12 sets of data before any meaningful analysis can be carried out. By contrast, if excessive data are collected it slows down collection and analysis processes, and leads to an over complex M&T system. In some situations data from various sources may be incompatible, making comparisons very difficult. For example, utility invoices may be collected which cover different time periods. In addition, errors may occur when meters are read incorrectly and also when readings are incorrectly logged.

7.3.1 Data from invoices

In most applications, energy invoices are the primary source of energy data. Data collection from invoices involves collating various utility and fuel bills, extracting relevant information and inputting data into a computer. Problems arise when invoices are misplaced. It is therefore important to establish good collection procedures which ensure that good quality data are entered into the computer system.

Although an extremely important source of data, in many larger process industries the use of monthly utility invoices is considered inadequate. This is because the data provided by the invoices are of insufficient detail and also because the data collection process is too slow, with the potential for excessive energy wastage. Therefore, in situations where invoice data alone are considered insufficient, it is necessary to establish more detailed data collection systems.

7.3.2 Data from meters

Meter reading provides another useful source of energy data. Reading of meters should in theory be a relatively simple task, but unfortunately in practice a number of problems can arise. Some of the common faults which may arise are as follows [1]:

- Digits may be recorded in the wrong order.
- Too many digits may be inserted in the recorded reading.
- The recorded reading may be wrong by a factor of 10 or even 100.
- Readings may be lost after being recorded.
- The wrong meter may be read.
- Meters may not be read at all.
- Poor writing may be used when recording readings.

Not surprisingly, poorly recorded meter readings can result in much wasted time and effort. Therefore meter readings should always be validated. Validation checks should include [1]:

- Checking that the correct number of digits is recorded.
- Checking that current readings are higher than previous readings.
- Checking that readings are within predicted energy consumption bands.
- Checking the date of meter readings.

Should a meter reading fail one of the above tests, the most likely source of the discrepancy will be operator error, although the fault cannot always be attributed to the operator. For example, a meter may have been changed, or, for some unknown reason, there may have been a period of

abnormally high energy consumption. It is therefore important to establish a robust data validation system, one that alerts the system user as soon as a potential error is detected. Validation can
take place either when readings are input, or at the initial stages of the data analysis.

The manual reading of meters and the writing down of digits by hand is a time consuming
process which is prone to mistakes. With the advent of smart metering systems, there are a
variety of alternatives to the manual approach. For example, data capture units can be employed.
These are small portable computer units, into which meter readings can be directly entered. The
data collected in this way can later be downloaded to a larger personal computer for analysis.
Data capture units can also be programmed to validate data as it is input. Data capture units thus
improve the quality of manual meter reading, but they still require an operator to visit all the
relevant meters, which is a labour intensive process. By using pulsed output meters it is possible
to carry out remote meter reading automatically. Pulsed output meters send out an electronic
signal which can be automatically monitored by an M&T computer. Many BMS also have the
capability to monitor energy consumption and data can be downloaded to an M&T computer
either as 'ASCII' or spreadsheet files.

In situations where existing metering is unable to provide enough detailed information on
energy consumption, it may be necessary to install additional sub-metering (see Chapter 5). This
should improve the overall quality of any M&T programme and result in increased energy
savings. However, additional metering can be expensive and can also lead to a substantial
increase in the quantity data which must be collected and processed. Therefore any additional
metering must be justified, to ensure that potential energy cost savings are not outweighed by the
installation costs.

7.4 Energy targets

At the outset of any M&T programme it is important to set energy targets. Initially, these
should be the *standard energy performances* which have been established for the respective
EACs [2]. *Standard performance* can be determined through analysis of past energy
consumption data (see Chapter 4). If these data do not exist, then it will be necessary to
undertake an auditing process to establish credible *standard energy performances* for the
respective EACs.

Although the standard energy performance is generally used as an initial target, subsequent
energy targets should represent improvements on standard performance. One way of establishing
an improved target is to plot a *best-fit* straight line through the lower edge of the points on a scatter
diagram (see Figure 7.1). At first sight this might appear as a strange thing to do, but the points
on the lower edge of the scatter diagram represent the historical achievement of least energy
consumption and therefore should be an achievable target. In other words a straight line through
the lower points represents what can actually be achieved in practice and makes an ideal energy
target.

Energy targets should be reappraised on a regular basis. This can be done by defining the best
historical performance as the target in a similar way to that described above, or by basing the
target on an agreed action plan that is designed to achieve energy savings. Both these methods
have the advantage of being based on real data and so should be achievable. A more arbitrary,
and possibly inferior, approach is to set targets based on a percentage improvement on the
current energy performance. No matter the approach chosen, all targets should be realistic and
achievable otherwise they will lose credibility. They should also be reviewed regularly, in order
to maintain pressure to reduce energy consumption.

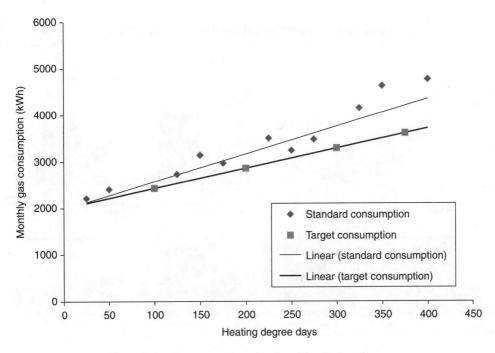

Figure 7.1   Target curve based on best historical performance

7.5 Reporting

One of the major outputs of any M&T programme is the production of energy management reports. These reports perform the vital role of communicating key information to senior and operational managers, and are therefore the means by which action is initiated within an organization. In order to ensure that prompt action is taken to minimize wasteful practices, reports should be as simple as possible and should highlight those areas in which energy wastage is occurring. Reports should be published regularly so that energy wasteful practices are identified quickly and not allowed to persist for too long. Reports should be succinct, and conform to a standard format which should be generated automatically by a computer. This minimizes preparation time, and also familiarizes managers with the information being communicated. Figure 7.2 shows an example of a weekly electricity consumption report produced by an M&T computer program.

Most M&T programmes require reports to be published weekly or monthly. Monthly reports are usually applicable to large organizations with many sites, with weekly reports being more suitable to complex high energy consuming facilities. In applications where energy consumption is particularly high, reports may be produced daily. If the reporting period is too long, energy may well be needlessly wasted before managers are notified of the problem and remedial action is taken. Yet, if the reporting period is too short this will lead to an over-complex M&T system in which too much irrelevant information requires consideration.

The primary purpose of energy management reports is to communicate effectively with senior and operational managers. They should therefore be tailored to suit the needs of their readers, with different managers within organizations requiring different levels of report. Operational managers may need weekly reports, whereas senior management may only require

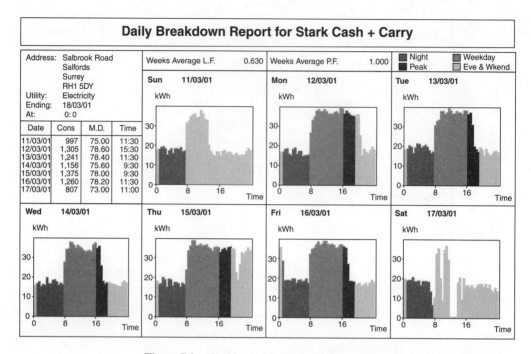

Figure 7.2   Weekly electricity consumption report[3]

a quarterly review. Figure 7.3 illustrates the relationship between reporting frequency and managerial status.

One big disadvantage of producing a large number of regular reports is that they can swamp operational managers with what may appear to be irrelevant information. One good way to get around this problem is to adopt a *reporting by exception* system, in which reports are only generated when energy performance falls outside certain predetermined limits. This system has the great advantage that managers only receive reports when energy performance is either poor or very good. In addition, everyone involved in the reporting process benefits from a reduced workload.

Level	Report frequency			
Senior management	Annual	Quarterly		
Department head	Annual		Monthly	
EAC manager	Annual			Weekly
Energy manager	Annual	Quarterly	Monthly	Weekly

Figure 7.3   Relationship between managerial status and report frequency. Crown copyright is reproduced with the permission of the Controller of Her Majesty's Stationery Office and the Queen's Printer for Scotland

Table 7.1   League table of shop electrical consumption

Shop	Floor area (m²)	Electrical units consumed (kWh)	Electricity cost (£)	Electrical units per m² (kWh/m²)	Average cost per unit (p/kWh)
Crewe	6 560	91 906	4374.72	14.01	4.76
Chester	12 000	149 845	6967.79	12.49	4.65
Leeds	13 600	169 400	6860.70	12.46	4.05
Bingley	6 070	73 677	5046.87	12.14	6.85
Macclesfield	6 460	74 823	3165.01	11.58	4.23
Huddersfield	9 470	108 640	5442.86	11.47	5.01
Stockport	10 800	123 892	4943.29	11.47	3.99
Doncaster	12 500	132 345	5426.15	10.59	4.10
Ashbourne	5 780	59 534	2393.27	10.30	4.02
Burnley	7 540	74 943	3050.18	9.94	4.07
Halifax	7 800	72 513	2806.25	9.30	3.87
Stoke	8 670	75 342	3337.65	8.69	4.43

7.6 Reporting techniques

There are a number of reporting techniques which can help energy managers communicate effectively with operational managers. These include the production of league tables and a variety of graphical techniques.

7.6.1 League tables

League tables can be a particularly effective way of conveying energy performance information. They are most useful when comparing a number of similar EACs. Example 7.1 illustrates this process.

Example 7.1

A retail chain has 12 shops. Table 7.1 shows a league table of the monthly electrical energy consumption and cost of all 12 shops.
From the league table, two issues become obvious:

- The shop at Crewe appears to have an abnormally high electricity consumption compared with similar sized shops at Bingley, Burnley and Macclesfield. Further investigation is therefore required to identify the reason for the high electricity consumption.
- The shop at Bingley is paying too much for its electricity, which implies that the tariff needs to be changed.

If an organization has a number of dissimilar EACs then direct comparison by league table is meaningless. However, league tables can still be usefully employed if the EACs are ranked on the basis of variation from target performance, as shown in Table 7.2.

From Table 7.2 it is clear that in energy consumption terms, the Geography Department is performing poorly, while the Mechanical Engineering Department is performing relatively well.

Table 7.2   League table of electrical consumption for various university departments

EAC	Electrical units consumed (kWh)	Target use (kWh)	Electricity cost (£)	Excess above target (kWh)	Variation from target (%)
Geography dept	11 780	10 484	565.44	+1296	+12.36
Electrical eng. dept	40 056	37 653	1922.69	+2403	+6.38
Civil eng. dept	50 834	48 801	2440.03	+2033	+4.17
Law dept	8 893	8 982	426.86	−89	−0.99
Chemistry dept	180 567	183 817	8667.22	−3250	−1.77
Mechanical eng. dept	72 004	75 604	3456.19	−3600	−4.76

7.6.2 Graphical techniques

Graphical techniques, if used correctly, can be an excellent tool for communicating information to managers. However, not all graphical techniques are easy to follow and some can be very misleading. For example, although the *scatter diagram* is an excellent analysis tool (see Chapter 4), it is a poor communication tool because it gives no historical record of energy consumption. Since it is unlikely that senior and operational managers will be familiar with energy analysis techniques, it is important that energy performance graphs be of a type that is easily understood by non-energy specialists.

There are a number of graphical techniques which can be used to represent historical energy consumption. Of particular importance are the *norm chart*, the *deviance chart*, and the *CUSUM chart*. The *norm chart* is a sequential plot of actual energy consumption overlaid on a plot of target consumption (see Figure 7.4). It is of little value as an analytical tool, but can be very

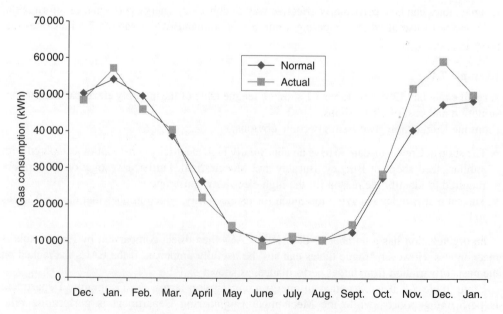

Figure 7.4   Norm chart for gas consumption

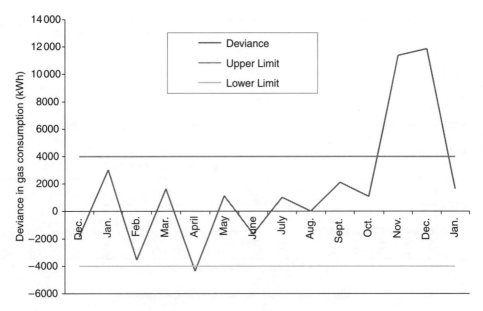

Figure 7.5   Deviance chart for gas consumption

useful for highlighting exceptions and communicating these to managers. Because *norm charts* represent a historical record of energy consumption, senior and operational managers find them relatively easy to understand.

Deviance charts plot the difference between target and actual energy consumption (see Figure 7.5). If, in any one month, energy consumption is above the target value, then the consumption is plotted as a positive value; by contrast a negative value is returned if actual consumption is lower than predicted. When producing a deviance chart it is useful to show on the graph limits of normal operation, since this helps to distinguish between normal back-ground 'noise' and serious deviations from the norm. Deviance charts are particularly good at highlighting problems, so that remedial action can be taken. They can also be used to initiate detailed *exception reports*.

The concept of the CUSUM is discussed in detail in Chapter 4. *CUSUM charts* plot the cumulative sum of the deviation of actual energy consumption from predicted or target consumption. As such they are a useful tool when identifying trends and diagnosing problems. They are also useful when assessing the impact of any remedial action which is taken. While *CUSUM charts* are a useful diagnostic tool, they can be difficult to understand if not fully explained. Care is necessary to ensure that managers understand what is being communicated in a *CUSUM chart*. Figure 7.6 shows a typical *CUSUM chart*.

7.7 Diagnosing changes in energy performance

CUSUM charts can be particularly useful when diagnosing why energy wastage is occurring. This is principally because they identify the date of any change in energy performance. It can be particularly helpful to know when a problem first occurred, as this helps to pin-point the prob-lem; further analysis can then be undertaken to determine its cause. The diagnostic use of CUSUM is illustrated in Example 7.2.

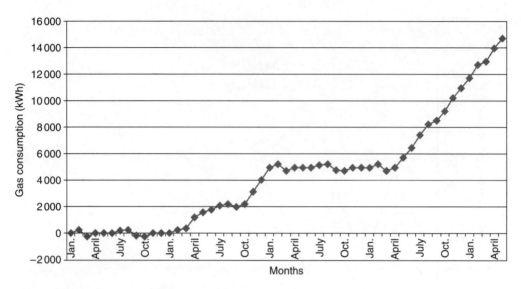

Figure 7.6   CUSUM chart for gas consumption

Example 7.2

Figure 7.6 shows a CUSUM plot of gas consumption for a residential building for the period January 1997 to May 2001. The CUSUM plot shows that for the period January 1997 to February 1998 the actual energy consumption conformed to *standard energy performance*. However, around March 1998 something happened to change the energy performance of the building dramatically for the worse and poor performance continued until February 1999 when the problem was rectified. In May 2000 the energy consumption again took a dramatic turn for the worse and this continued unresolved until the end of the analysis period.

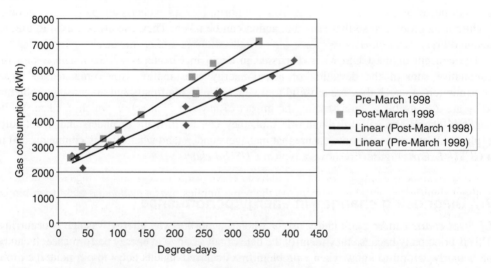

Figure 7.7   Performance lines pre- and post- March 1998, indicating that reduced performance is due to a weather-related matter

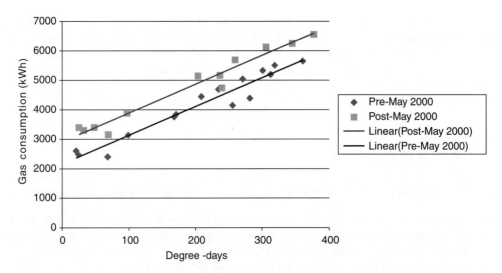

Figure 7.8   Performance lines pre and post May 2000, indicating that reduced performance is due to an increased base load

In order to determine why the change in performance occurred in March 1998, regression analysis needs to be carried out for periods 'pre and post' the event. Figure 7.7 shows the results of the regression analysis.

From Figure 7.7 it is clear that there is a change in the slope of the *best-fit* straight-line curve which indicates that the increase in energy consumption is weather-related. The point at which both lines intersect the 'y'-axis is the same, indicating that base load consumption has not increased. Possible explanations for the increase in weather-related performance, could be:

- An increase in the space temperature control setting.
- An increase in the ventilation rate. If the building had a central ducted warm air system, this situation could occur if the fresh air mixing dampers on the main air handling units were incorrectly set.

Now consider the increase in energy consumption which occurred in May 2000. Figure 7.8 shows the regression analysis plot for the periods before and after this date. It can be seen that the slope of the two 'best fit' straight lines is the same. This indicates that the increase in energy consumption was not a weather-related issue, but rather caused by a significant increase in the base load. This could have been caused by incorrect setting of the thermostat on the hot water storage cylinders in the building.

7.8 Waste avoidance

Waste avoidance is a simple concept, which can enable large energy cost savings to be achieved without significant capital investment. As the name implies, *waste avoidance* seeks to minimize avoidable wasted energy. *Waste avoidance* programmes should be quick and inexpensive to implement, since their specific objective is cut out waste that can be easily avoided.

Unlike materials wastage, which is usually relatively easy to detect, energy wastage is not at all easy to spot and identifying it usually requires considerable detective work. It is often the case that large quantities of energy can be wasted over long periods of time, without anybody being aware that wastage is occurring. Consider the case of electric frost protection heaters. These are switched on in winter to protect control panels and other items of equipment from frost damage. Although they may be switched on in winter, it is often the case that they are not switched off when the cold weather has finished. As a result, they remain switched on all year round without anyone noticing. It is not difficult to see that if something like this is detected early enough, large energy savings can be made. Early diagnosis is essential, and a waste avoidance programme utilizing many of the techniques already described in this chapter will facilitate this.

7.9 Causes of avoidable waste

Although facilities and organizations may vary widely, many of the causes of avoidable waste are generic and common to most applications. These generic causes include the following [2]:

- Frost protection devices frequently remain in operation for long periods when there is no risk of frost damage. Under-surface heaters often remain on throughout the spring and summer without any perceptible effect. Similarly, pre-heating coils in air handling units often operate all the time, without the system operators being aware of the problem.
- The failure of switches can often cause major energy wastage. Switches are often employed to automatically turn equipment on and off. If for any reason a switch fails in the 'on' position, then the equipment will run continuously, often unnoticed, for a considerable time.
- Time controls can often lead to excessive energy consumption. Time switches are designed to turn off equipment after a pre-set period of operation. If they fail or are overridden, or indeed, if they are simply incorrectly set, it can result in extended plant operation and much excess energy being needlessly consumed.
- Lack of suitable control often results in unnecessarily high energy consumption. Items of equipment such as luminaires and ventilation fans often have no controls other than a manual switch. Such items tend to be left on continuously. Remember, staff and operatives, although quick to switch on items of equipment, often forget to switch them off when they are not needed.
- Leaks of water or steam from pipes often go unnoticed for long periods, since they usually occur out of sight.
- Control valves and dampers, if not correctly monitored and controlled, can result in excessive energy wastage. Air handling systems often employ modulating dampers to vary the quantities of fresh and recirculated air. However, if a damper remains in the 'full fresh air' position for long periods during cold weather, much energy will be wasted in heating up excessive quantities of cold air.
- With air conditioning systems, much energy can be wasted due to poorly set controls. If the cooling 'set-point' is set below the heating 'set-point', then both the cooling and heating coils will operate at the same time and 'fight' each other. This can occur without the building occupants being aware and always leads to excessive energy consumption.

It should be noted that most of the causes of the energy wastage listed above are due to either incorrect control setting, control failure, or the use of insufficient controls. These are failures which can be rectified at relatively little cost.

A good way to detect and solve energy wastage problems is to create checklists which identify common problem areas. Checklists can also assist in the prioritization of energy conservation measures. They should commence with the more obvious and simple energy saving measures and progress towards those which are more obscure and which may demand capital expenditure.

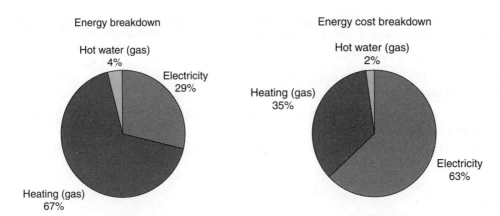

Figure 7.9   Factory unit: annual energy and cost breakdown

7.10 Prioritizing

Potential energy efficiency measures should be prioritized. This involves identifying and quantifying energy costs, and highlighting those measures which offer the greatest potential savings.
History is littered with schemes which were embarked upon because they were fashionable or
perceived as 'sexy' and which in the long run proved to be ineffective, ultimately becoming
'follies'. Such schemes usually come about because someone at the planning stage did not
correctly quantify the potential gains. Very often it is the mundane measures which yield the best
returns.

One way to prioritize action areas is to produce league tables as discussed in Section 7.5.1.
This enables those EACs requiring most urgent investigation to be identified. When prioritizing
measures it is important to assess both the energy consumption and the energy cost. Consider the
case of a factory building with a gas fired heating system, the annual energy and cost breakdowns
of which are shown in Figure 7.9.

It can be seen from Figure 7.9 that although electrical energy use is 29 per cent of total energy
use, the cost of the electrical energy represents 63 per cent of the total energy bill, 1.7 times the
expenditure on gas. Consequently, priority should be given to reducing electrical consumption
rather than gas consumption.

References

1. Computer-aided monitoring and targeting for industry, Good Practice Guide 31 (1995).
   Department of the Environment.
2. Waste avoidance methods, Fuel Efficiency Booklet 13 (1995). Department of the
   Environment.
3. Stark energy information systems (2001). http://www.stark.co.uk/, 16 July.

Bibliography

Computer-aided monitoring and targeting for industry, Good Practice Guide 31. (1995).
Department of the Environment.

Eastop, T. D. and Croft, D. R. (1990). *Energy efficiency for engineers and technologists*, (Chapter 9). Longman Scientific & Technical.

Energy audits and surveys (1991). CIBSE Applications Manual AM5.

Moss, K. J. (1997). *Energy Management and Operating Costs in Buildings*, (Chapter 10). E & F. N Spon.

Monitoring and targeting in large manufacturing companies, Good Practice Guide 91. (1994). Department of the Environment.

Waste avoidance methods, Fuel Efficiency Booklet 13 (1995). Department of the Environment.

8

Energy efficient heating

This chapter outlines good practice in the design and operation of building heating systems. It includes a discussion of the types of heating system that should be selected for particular applications. In addition, the concept of building heat loss is discussed, and methodologies presented for predicting heating energy costs and optimizing plant utilization.

8.1 Introduction

For those who live in temperate, cool or cold climates, the provision of adequate heating is essential for large parts of the year in order to maintain comfort and good health. While some may consider good 'central heating' to be a luxury, a recent epidemiological study undertaken in several European countries revealed that death rates amongst those over 50 years of age are proportional to the degree of coldness experienced [1]. In fact, in England and Wales there is a 2% increase in mortality for every 1 °C below an outside temperature of 19 °C [2]. As a result, weather related mortality rates in the UK are amongst the highest in Europe, with an estimated 40 000 additional deaths occurring during the winter months each year [1]. One major contributory factor to this sad state of affairs is the poor state of much of the housing stock in the UK [1, 3].

In addition to preserving the good health of building occupants, adequate heating of public, commercial and industrial buildings is essential to promote efficient and productive work practices. If people are cold and uncomfortable, they will not perform well and so output will fall. Clearly it is false economy to save a little on fuel costs and lose much in productivity.

Although a building may be heated to a comfortable level, it does not always mean that it is efficiently heated. In many buildings large quantities of heat energy are wasted daily due to a combination of poor design and poor operating and maintenance practices. Broadly speaking, heat can be wasted by any combination of the following:

- Poorly designed heating systems, which are often wrongly selected for particular applications.
- Poorly designed and insulated building envelopes.
- Poorly insulated heating systems.
- Poorly commissioned and maintained boiler plant.
- Poor controls.
- Poor operating practices.

To understand the reasons why heat is so easily wasted in buildings, it is worth considering the case of an old poorly maintained church hall with a high roof, which is heated by an old low temperature hot water (LTHW) radiator system, served by an oil-fired boiler. Let us assume that the building is used for meetings and social events, mainly in the evenings and at weekends. Without going into too much detail, there are a number of possible ways in which the heating system might be considered to be inefficient:

- The building envelope is probably poorly insulated so heat will be easily lost to the outside by conduction, especially on cold winter nights.
- The building envelope is probably not very well sealed and so large amounts of the warm heated air will be lost through cracks around doors and window frames.
- In the church hall the warm air heated by the radiators will rise up into the high roof space where it will stratify. Consequently, any pigeons that might be in the roof space will be warm, while the occupants at floor level will probably feel cold and uncomfortable.
- The pipework from the boiler to the radiators may not be very well insulated. This will result in major heat losses if the pipes run through unheated spaces, such as floor voids or in the boiler room.
- The boiler will probably be poorly maintained, with the result that some of the heat meant to warm up the water will be lost up the flue with the combustion gases.
- Since the system is old, it is probable that its controls will be inadequate. For example, the system might have a time clock which turns the boiler and pump on at the same time each day, whether or not the hall is occupied. Alternatively, the thermostat in the room space may be set at too high a temperature, so that the hall over-heats and the occupants become uncomfortable. It is not uncommon for the occupants of buildings to open windows rather than turn down thermostats. After all, it is often the case that the occupants of a building are not the ones paying the fuel bills.

The illustration of the poorly heated church hall, demonstrates four important points:

1 It is essential to take a holistic view when designing a heating system. The building fabric and the mechanical heating system are equally important.
2 It is important to select the correct heating system for the particular application.
3 It is important to have an adequate control system.
4 It is no use designing and installing an excellent heating system, if the building operatives and occupants do not understand how to use it correctly.

8.2 Thermal comfort

When an individual is in a room, his/her vital organs are maintained at a temperature of 37.2 °C through a complex set of heat transfer mechanisms. The person will lose heat to the surrounding air by convection, and to any cold surfaces within the room space by radiation. However, hot surfaces within the room will cause the person to gain heat by radiation. Consider a room in which the air is maintained at 21 °C by wall-mounted radiators which have a surface temperature of 70 °C. The occupants of the room will lose heat by convection to the air, because the surface of their clothes will be at approximately 30 °C. They will also lose heat by radiation to many of the cool surfaces in the room, most of which will be at a temperature of less than 21 °C. At the same time they will gain heat by radiation from the hot radiators. The occupants will also lose heat by evaporation through exhalation and perspiration. A small amount of heat will also be lost through their feet by conduction. Consequently, a complex heat balance is set-up, which maintains the core temperature of the occupant's bodies at 37.2 °C. If for any reason a person becomes hot or cold, in other words uncomfortable, then their body will take involuntary action in order to maintain the core temperature. One such mechanism is perspiration, which increases under warm conditions, so that increased evaporative cooling occurs. If a person is uncomfortable, he/she can also take voluntary steps to rectify the situation. For example extra/excess clothing can be worn or removed. Equation 8.1 expresses the heat balance between the human body and the surrounding environment [4]. It should be noted that the term for conduction has been omitted from eqn (8.1) because in most normal situations it is negligible.

$$M - W = Q_e \pm Q_c \pm Q_r + S \qquad (8.1)$$

where M is the metabolic rate (W), W is the rate at which energy is expended in mechanical work (W), Q_c is the rate of heat transfer by convection (W), Q_r is the rate of heat transfer by radiation (W), Q_e is the rate of heat loss by evaporation (W), and S is the rate at which heat is stored in the body (W).

Food that is digested by the body is converted into energy. Some of this energy is used to perform mechanical work, but most of it produces heat. In fact, under normal conditions, more heat is produced by the body than it actually requires to maintain its core temperature. Therefore heat is lost from the body by convection, radiation and evaporation. The human body can also gain heat by convection and radiation under hot conditions. It is however, impossible to gain heat by evaporation, since this is always a cooling mechanism. If a person performs exercises or mechanical work, then the metabolic rate increases and the body is required to reject more heat, otherwise it will over heat. Consequently, in rooms such as gymnasia, where vigorous exercise is performed, it is necessary to have a lower room air temperature to compensate for the increased heat production rate of the human body. Table 8.1 shows heat outputs for a range of work rates, together with recommended room temperatures.

The thermal comfort of building occupants can be affected by personal and environmental factors. Personal factors can be defined as those variables which are directly connected with the individual. These include:

- Activity: The higher the level of personal activity, the greater the heat output.
- Clothing: The greater the amount of clothing worn, the lower the heat loss.
- Age: The metabolic rate decreases with age.
- Gender: The resting metabolic rate of women is approximately 10% lower than that of men.
- Health: Illness affects the ability of the body to maintain its core temperature at 37.2 °C.

These personal comfort factors should always be considered when designing heating systems for particular applications. For example, because the metabolic rate of the elderly is lower than that of younger people, it is important to maintain air temperatures at a higher than normal level in sheltered

Table 8.1 Sensible heat outputs and recommended dry resultant temperatures [5]

Type of work	Typical application	Required room dry resultant temperature (°C)	Sensible heat output per person (W)
Light work	Office	20.0	100
Walking slowly	Bank	20.0	110
Light bench work	Factory	16.0	150*
Heavy work	Factory	13.0	200*

* Assumes a dry-bulb temperature of 15 °C.

accommodation for the elderly. Similarly, when designing the foyer of a railway station, where passengers in transit may be wearing outdoor winter clothes, it is advisable to maintain the air temperature at a lower temperature than, say an office, where indoor clothing would normally be worn.

The environmental parameters which influence thermal comfort are: air temperature, mean radiant temperature, relative humidity and air velocity. These parameters, in different ways, influence heat transfer by convection, radiation and evaporation. Convective heat transfer is influenced by air temperature and air velocity. It takes place continually, although in most buildings it is imperceptible because air speeds are low (i.e. below 0.1 m/s). To be perceptible, air speeds must exceed approximately 0.2 m/s. Radiant heat transfer occurs between the skin/clothes and those surfaces 'seen' by the human body. It is therefore heavily influenced by the *mean radiant temperature* (t_r) of the surfaces within rooms. For cuboid shaped rooms, the approximate *mean radiant temperature* at the centre of the room can be determined by:

$$t_r = \frac{a_1 \cdot t_1 + a_2 \cdot t_2 + a_3 \cdot t_3 + \cdots + a_n \cdot t_n}{a_1 + a_2 + a_3 + \cdots + a_n} \tag{8.2}$$

where t_r is the mean radiant temperature (°C), $a_1, a_2, \ldots,$ are the room component surface areas (m²), and $t_1, t_2, \ldots,$ are the room component surface temperatures (m²).

Evaporative heat loss is governed by the relative humidity of air, and air velocity. If conditions are very humid, as in tropical countries, then evaporative heat losses will be low, since any perspiration produced is unable to evaporate. If, however, the air is dry then evaporation readily takes place and the body is cooled. Evaporative cooling becomes an important heat transfer mechanism at air temperatures above 25 °C. At air temperatures above 29 °C almost all heat is lost from the body by evaporation [6].

Thermal comfort is determined by a number of environmental factors, particularly *mean radiant temperature*, air temperature and air velocity. It is thus important that these factors are considered when designing buildings. The concept of *dry resultant temperature* (t_{res}) was developed to make allowances for these considerations. The *dry resultant temperature* is defined as:

$$t_{res} = \frac{t_r + t_a \cdot \sqrt{10v_a}}{1 + \sqrt{10v_a}} \tag{8.3}$$

where v_a is the air velocity (m/s), and t_a is the air temperature (°C).

However, in most buildings, because the air velocity is below 0.1 m/s, eqn (8.3) can be simplified to:

$$t_{res} = 0.5\, t_r + 0.5\, t_a \tag{8.4}$$

Equation 8.4 tells us that *mean radiant temperature* is as important as air temperature when maintaining a comfortable environment within buildings. This explains why on returning to an unheated house after some days of absence, the rooms will feel cold and uncomfortable, even though the heating is on and the air up to temperature. The air temperature may be acceptable, but the dry resultant temperature will still be low because the fabric of the building is still cold.

8.3 Building heat loss

When a building is heated to a steady internal temperature, an equilibrium is established in which the heat power into the building equals the rate at which heat is lost from the building. Thus to maintain a comfortable internal environment, the output from any heating system must be equal to or greater than the combined effect of the heat loss through the building fabric and the ventilation heat losses. Fabric heat losses are those which occur primarily by conduction through walls, windows, floors and roofs. Ventilation losses are the convective heat losses which occur when warm air is lost from a building and replaced by cold air. A good approximation of the fabric heat losses can be determined using the generic equation:

$$Q_f = U \times A \times (t_{ai} - t_{ao})$$ (8.5)

where Q_f is the fabric heat loss rate (W), U is the thermal transmittance (U value) (W/m²K), A is the area (m²), t_{ai} is the internal air temperature (°C), and t_{ao} is the external air temperature (°C).

Similarly, the ventilation heat loss can be determined using:

$$Q_v = 0.333 \times n \times V \times (t_{ai} - t_{ao})$$ (8.6)

where Q_v is the ventilation heat loss rate (W), n is the ventilation rate (air changes per hour), and V is the volume (m³).

From eqns (8.5) and (8.6) it can be seen that both the fabric and the ventilation heat loss rates are directly proportional to the difference between the internal and the external air temperatures. Since the internal air temperature should be maintained at a constant level during the winter, the heat loss therefore varies with the outside air temperature. Boilers and heat emitters should be sized for the 'worst-case' scenario (i.e. a very cold day) and be capable of maintaining the internal design temperature under this extreme weather condition. When operating under less extreme conditions (i.e. under part-load conditions), the output of the heating system can be reduced by using controls. Table 8.2 gives some sample winter external design conditions for various parts of the world.

The simplest way to reduce the energy consumed by a heating system is to create a building envelope which is well insulated and in which ventilation rates are controlled to the minimum required for healthy living. Achieving this usually involves care and attention during both the design and the construction stages, together with increased capital outlay. Not surprisingly, many speculative builders have little incentive to construct energy efficient building envelopes, since they usually do not pay the fuel bills. Therefore most countries have building regulations, of varying degrees of rigour, to force developers to conform to certain minimum thermal insulation standards. Usually these standards are expressed in terms of maximum permissible U values for glazing, walls, floors and roofs.

8.3.1 U values

Thermal insulation standards are usually expressed in terms of U values. The U value, or thermal transmittance, is a measure of the overall rate of heat transfer, under standard conditions, through a particular section of construction. It has units W/m²K and, as can be seen from eqn (8.7), is the inverse of thermal resistance. The lower the U value of an item of construction, the better its thermal insulation performance.

$$U = \frac{1}{\Sigma R}$$ (8.7)

where ΣR is the total thermal resistance of the construction (m²K/W).

The thermal resistance of each component layer of any construction can be determined by using eqn (8.8).

$$R = \frac{l}{\lambda}$$ (8.8)

Table 8.2   External winter design temperatures for sample cities around the world [7]

Country	City	Winter external design temperature (°C)
Australia	Perth	6
	Sidney	6
Belgium	Brussels	−7
China	Shanghai	−3
France	Lyon	−10
	Paris	−4
Germany	Berlin	−11
	Hamburg	−9
	Munich	−13
Italy	Milan	−6
	Naples	2
	Rome	1
India	New Delhi	4
Japan	Tokyo	−2
New Zealand	Christchurch	−1
	Wellington	3
Norway	Oslo	−16
Spain	Barcelona	2
	Madrid	−2
Sweden	Stockholm	−13
United Kingdom	Birmingham	−3
	Glasgow	−2
	London	−2
	Manchester	−2.5
USA	Chicago	−20
	Dallas	−6
	Kansas City	−14
	Los Angeles	4
	Miami	8
	New Orleans	1
	New York	−9
	San Francisco	3
	Seattle	−9
	Washington DC	−8

where R is the thermal resistance of layer (m^2K/W), l is the thickness of layer (m), λ is the thermal conductivity (W/mK).

The total thermal resistance of a construction is the summation of the resistances of all the individual layers, plus the resistances to heat transfer of any surfaces. Table 8.3 gives typical thermal conductivity values for a variety of building materials.

Figure 8.1 illustrates the heat transfer mechanisms that take place in a typical external wall with an air cavity. While the heat transfer through the solid parts of any construction is by conduction, it is radiation and convection that control the heat transfer at the surfaces. The resistance to heat transfer of a surface can be calculated using eqn (8.9), which takes into consideration the convective and radiative heat transfer coefficients, and the emissivity of the surface.

Table 8.3   Thermal conductivity of various materials [8]

Material	Density (kg/m³)	Thermal conductivity (W/mK)
Brickwork (outer leaf)	1700	0.84
Brickwork (inner leaf)	1700	0.62
Cast concrete (dense)	2100	1.40
Cast concrete (lightweight)	1200	0.38
Concrete block (heavyweight)	2300	1.63
Concrete block (medium weight)	1400	0.51
Concrete block (lightweight)	600	0.19
Fibreboard	300	0.06
Plasterboard	950	0.16
Plaster (dense)	1300	0.50
Plaster (lightweight)	600	0.16
External rendering	1300	0.50
Screed	1200	0.41
Asphalt	1700	0.50
Tile	1900	0.84
Wood-wool slab	650	0.14
Expanded polystyrene	25	0.035
Glass fibre slab	25	0.035
Phenolic foam	30	0.040

$$R_s = \frac{1}{h_c + (E \times h_r)} \tag{8.9}$$

where R_s is the surface resistance (m²K/W), h_c is the convective heat transfer coefficient (W/m²K), h_r is the radiative heat transfer coefficient (W/m²K), and E is the emissivity factor.

The convective heat transfer coefficient is the rate at which heat is transferred to or from a 1 m² surface by convection, per 1 °C temperature difference between the surface and the neighbouring fluid (i.e. air). The radiative heat transfer coefficient is the rate of radiant heat transfer to or from a 1 m² surface of a *black body* divided by the difference between the mean temperatures of the radiating surface and that of the surrounding surfaces. The term *black body* refers to a body which

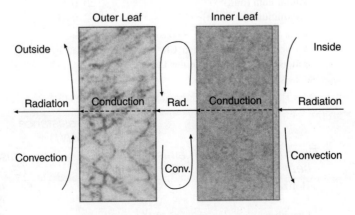

Figure 8.1   Heat transfer through a cavity wall

absorbs all energy incident on it at every wavelength and conversely emits energy at every wave-length. The emissive power of a black body can be calculated using eqn (8.10).

$$E_b = \sigma \times T^4 \tag{8.10}$$

where E_b is the emissive power (W/m^2), σ is the Stefan–Boltzmann constant (i.e. 5.67×10^{-8} W/m^{2}K^4), and T is the absolute temperature of the black body (K).

Of course, black bodies are theoretical and building materials do not emit and absorb energy at all wavelengths and are therefore non-black bodies. The term *emissivity* is used to describe the ratio of the emissive power of a non-black body to that exhibited by a black body. The emissivity of a black body is 1, while that of all other bodies will be less than one. Dull surfaces, of the type exhibited by most building materials, have a high emissivity, while shiny metallic surfaces have a low emissivity. High emissivity surfaces are good emitters and receivers of radiation, while low emissivity surfaces are not. Table 8.4 gives examples of the emissivity values for typical building materials.

The emissivity factor (E) referred to in eqn (8.9) allows for the emissivity and geometrical relationship of both the emitting and the receiving surfaces. The emissivity factor varies with the specific geometrical arrangement of the surfaces involved, but for most building applications it can be taken as being approximately 0.9.

The thermal resistance of any surface is strongly influenced by air velocity. However, for most building applications, room air velocities are not greater than 0.1 m/s and so it is possible to determine typical values for a variety of surfaces (as shown in Table 8.5). Table 8.5 presents typical internal surface resistances for both high and low emissivity surfaces.

The resistance of external building surfaces is heavily influenced by wind speed. Table 8.6 gives typical external surface resistances under sheltered, normal and severe conditions.

Unventilated air cavities in walls and roofs offer resistance to heat transfer. For cavities in walls the thermal resistance increases with thickness up to approximately 25 mm. Thereafter the

Table 8.4   Typical values of emissivity

Surface	Emissivity
Black body	1
Black (non-metallic)	0.90–0.98
Concrete	0.85–0.95
White paint	0.85–0.95
Aluminium (dull)	0.20–0.30
Aluminium (polished)	0.02–0.05

From Smith, Phillips and Sweeney (1987) *Environmental Science*, © Longman Group UK ltd (1987), reprinted by permission of Pearson Education Ltd.

Table 8.5   Typical internal surface resistances [8]

Building element	Direction of heat flow	Surface resistance (m^{2}K/W)	
		High emissivity factor (E = 0.97)	Low emissivity factor (E = 0.05)
Walls	Horizontal	0.12	0.30
Ceiling or roofs	Upward	0.10	0.22
Ceiling or floors	Downward	0.14	0.55

Table 8.6   Typical external surface resistances [8]

Building element	Emissivity of surface	Surface resistance (m²K/W)		
		Sheltered	Normal	Severe
Wall	High	0.08	0.06	0.03
	Low	0.11	0.07	0.03
Roof	High	0.07	0.04	0.02
	Low	0.09	0.05	0.02

Sheltered: Applies up to the third floor in city centres;
Normal: Applies to most suburban and rural areas, and from the fourth to the eighth floors on tall buildings in city centres; and
Severe: Applies to coastal and hill sites, above the fifth floor in suburban and rural districts and above the ninth in city centres.

thermal resistance of the air cavity is virtually constant despite any further increase in thickness. Heat transfer across an air cavity is by convection and radiation. Thus the emissivity of the cavity surfaces significantly influences the heat transfer. Table 8.7 gives typical thermal resistances for a variety of air cavities.

Table 8.7   Typical resistances of unventilated air cavities [8]

Cavity thickness	Emissivity of surface	Thermal resistance (m²K/W) for stated heat flow direction		
		Horizontal	Upward	Downward
5 mm	High	0.10	0.10	0.10
	Low	0.18	0.18	0.18
25 mm or greater	High	0.18	0.17	0.22
	Low	0.35	0.35	1.06

Example 8.1 illustrates how the above equation and data can be used to determine the U value of an external wall.

Example 8.1

An external wall has the following construction:

Element	Thickness (mm)	Thermal conductivity (W/mK)
Plaster	13	0.500
Concrete block (inner leaf)	100	0.200
Air cavity	50	n.a.
Brick (outer leaf)	102	0.840

Assuming that the air cavity has a thermal resistance of 0.18 m²K/W, the internal surface resistance is 0.123 m²K/W and the external surface resistance is 0.055 m²K/W, determine:

(i) The U value of the wall; and
(ii) The U value of the wall if the cavity is filled with insulating foam having a thermal conductivity of 0.036 W/mK.

Solution

Using the above data and eqn (8.8) it is possible to determine the total thermal resistance of the cavity wall as follows:

Element	Thickness (m)	Thermal conductivity (W/mK)	Thermal resistance (m²K/W)
Internal surface resistance	n.a.	n.a.	0.123
Plaster	0.013	0.500	0.026
Concrete block (inner leaf)	0.100	0.200	0.500
Air cavity	0.050	n.a.	0.180
Brick (outer leaf)	0.102	0.840	0.121
External surface resistance	n.a.	n.a.	0.055
		Total Resistance =	1.005

(i) Therefore:

$$U \text{ value (existing wall)} = \frac{1}{1.005} = 0.995 \text{ W/m}^2\text{K}$$

(ii) If the cavity of the wall is filled with foam, then:

$$\text{Thermal resistance of foam} = \frac{0.05}{0.036} = 1.389 \text{ m}^2\text{K/W}$$

However, the thermal resistance of 0.18 m²K/W for the air cavity no longer applies, therefore:

New thermal resistance of wall = 1.005 + 1.389 − 0.18 = 2.214 m²K/W

Therefore:

$$\text{New } U \text{ value} = \frac{1}{2.214} = 0.452 \text{ W/m}^2\text{K}$$

It can be seen from Example 8.1 that by filling the cavity with insulating foam, it has been possible to reduce the U value of the wall by 54.6%.

8.3.2 Heat loss calculations

If the U values for the various component parts of a building's fabric are known, then it is possible using eqns (8.5) and (8.6) to determine, with relative accuracy, the wintertime design day heat loss rate and thus ultimately size boiler plant and heat emitters. Example 8.2 illustrates how this calculation should be performed.

Example 8.2

The surface areas and U values of the elements of a building are as follows:

Element	Area (m²)	U value (W/m²K)
Floor	200	0.45
Roof	200	0.28
Single glazing	16	5.60
External doors	8	2.00
External walls	216	0.60

If the internal design temperature is 21 °C and the external design temperature is −1 °C, determine the design day heat loss rate. (Assume that the building experiences 3 air changes per hour and that its volume is 800 m³.)

Solution

Under the wintertime design condition, the temperature difference between the inside and outside is 22. Given this, and by applying eqn (8.5) (i.e. $Q_f = UA(t_{ai} - t_{ao})$), it is possible to calculate the heat loss rate through each component element of the building fabric.

Element	Area (m²)	U value (W/m²K)	Temperature difference (°C)	Heat loss (W)
Floor	200	0.45	22	1980.0
Roof	200	0.28	22	1232.0
Single glazing	16	5.60	22	1971.2
External doors	8	2.00	22	352.0
External walls	216	0.60	22	2851.2
			Fabric heat loss =	8386.4

By applying eqn (8.6) it is possible to determine the ventilation heat loss:

Ventilation heat loss = $0.3333 \times 3 \times 800 \times (21 - (-1)) = 17\ 600$ W

Now:

Total heat loss = Fabric loss + Ventilation loss

Therefore:

Total heat loss = $8386.4 + 17\ 600 = 25\ 986.4$ W

From an energy conservation point of view, one of the advantages of performing a design day heat loss calculation is that it gives an elemental breakdown of the relative heat losses from the building, enabling them to be quickly and easily evaluated. From Example 8.2 it can be seen that 1971.2 W is lost through the single glazing (i.e. 7.6% of the total heat loss). However, the ventilation loss of 17 600 W represents 67.7% of the total heat loss. Given this, it would be folly to install double glazing without first reducing the ventilation heat loss.

While the use of eqns (8.5) and (8.6) gives a relatively accurate value of the winter design day heat loss, it can be inaccurate, especially in applications where radiant heating is used. A superior model, which fully takes into account the radiant heat transfer which occurs within a room space, is described by eqns (8.11) and (8.12).

$$Q_f = F_1 \times \Sigma(AU) \times (t_c - t_{ao}) \tag{8.11}$$

and

$$Q_v = F_2 \times 0.333 \times N \times V \times (t_c - t_{ao}) \tag{8.12}$$

and

$$Q_p = Q_f + Q_v \tag{8.13}$$

where Q_p is the heating plant output (W), t_c is the dry resultant temperature in the centre of the room (°C), and F_1, F_2 are the characteristic temperature ratios.

The temperature ratios, F_1 and F_2, are defined as:

$$F_1 = \frac{(t_{ei} - t_{ao})}{(t_c - t_{ao})}$$
(8.14)

and

$$F_2 = \frac{(t_{ai} - t_{ao})}{(t_c - t_{ao})}$$
(8.15)

where t_{ei} is the internal environmental temperature (°C).

The internal environmental temperature is a theoretical construct which is used to calculate the radiative and convective heat transfer to the inside surface of an external wall from the other surfaces in a room. In temperate and hot climates, it can be defined as:

$$t_{ei} = \frac{1}{3} \cdot t_{ai} + \frac{2}{3} \cdot t_m$$
(8.16)

where t_m is the mean surface temperature of the room (°C).

The CIBSE publish tables of values for F_1 and F_2 for various heating systems [10]. Tables 8.8, 8.9 and 8.10 show values of F_1 and F_2 for a forced warm air system, a panel radiator system, and a high temperature radiant strip system. In each table values of F_1 and F_2 are presented against two variables, $\Sigma(AU)/\Sigma(A)$ and $NV/3\Sigma(A)$.

Once the values of F_1 and F_2 have been established for any system, it is possible to calculate the internal environmental temperature and the mean surface temperature of a room by using eqns (8.17) and (8.18).

$$t_{ei} = (F_1 \times (t_c - t_{ao})) + t_{ao}$$
(8.17)
$$t_{ai} = (F_2 \times (t_c - t_{ao})) + t_{ao}$$
(8.18)

Rearranging equation (8.16) gives:

$$t_m = \frac{3}{2} \cdot t_{ei} - \frac{1}{2} \cdot t_{ai}$$
(8.19)

Table 8.8 F_1 and F_2 values for 100% convective, 0% radiant heating (i.e. forced warm air heating)[10]

| | $\Sigma(AU)/\Sigma(A)$ | | | | | | | |
| | 0.2 | | 0.4 | | 0.6 | | 0.8 | |
$NV/3\Sigma(A)$	F_1	F_2	F_1	F_2	F_1	F_2	F_1	F_2
0.1	0.99	1.03	0.98	1.07	0.97	1.10	0.96	1.13
0.2	0.99	1.03	0.98	1.07	0.97	1.10	0.96	1.13
0.4	0.99	1.03	0.98	1.07	0.97	1.10	0.96	1.13
0.6	0.99	1.03	0.98	1.07	0.97	1.10	0.96	1.13
0.8	0.99	1.03	0.98	1.07	0.97	1.10	0.96	1.13
1.0	0.99	1.03	0.98	1.07	0.97	1.10	0.96	1.13
1.5	0.99	1.03	0.98	1.07	0.97	1.10	0.96	1.13
2.0	0.99	1.03	0.98	1.07	0.97	1.10	0.96	1.13
3.0	0.99	1.03	0.98	1.07	0.97	1.10	0.96	1.13
4.0	0.99	1.03	0.98	1.07	0.97	1.10	0.96	1.13

Table 8.9 F_1 and F_2 values for 70% convective, 30% radiant heating (i.e. panel radiators) [10]

| | $\Sigma(AU)/\Sigma(A)$ | | | | | | | |
| | 0.2 | | 0.4 | | 0.6 | | 0.8 | |
$NV/3\Sigma(A)$	F_1	F_2	F_1	F_2	F_1	F_2	F_1	F_2
0.1	1.00	1.01	0.99	1.03	0.98	1.05	0.98	1.06
0.2	1.00	1.00	0.99	1.02	0.99	1.04	0.98	1.06
0.4	1.00	0.99	1.00	1.01	0.99	1.02	0.99	1.04
0.6	1.01	0.97	1.00	0.99	1.00	1.01	0.99	1.03
0.8	1.01	0.96	1.01	0.98	1.00	1.00	1.00	1.01
1.0	1.02	0.95	1.01	0.96	1.01	0.98	1.00	1.00
1.5	1.03	0.92	1.02	0.93	1.02	0.95	1.01	0.97
2.0	1.04	0.89	1.03	0.90	1.03	0.92	1.02	0.93
3.0	1.06	0.83	1.05	0.85	1.05	0.86	1.04	0.88
4.0	1.07	0.78	1.07	0.80	1.06	0.81	1.06	0.83

Table 8.10 F_1 and F_2 values for 10% convective, 90% radiant heating (i.e. high temperature radiant systems) [10]

| | $\Sigma(AU)/\Sigma(A)$ | | | | | | | |
| | 0.2 | | 0.4 | | 0.6 | | 0.8 | |
$NV/3\Sigma(A)$	F_1	F_2	F_1	F_2	F_1	F_2	F_1	F_2
0.1	1.01	0.97	1.02	0.95	1.02	0.94	1.02	0.93
0.2	1.02	0.95	1.02	0.93	1.03	0.92	1.03	0.91
0.4	1.03	0.91	1.03	0.90	1.04	0.88	1.04	0.87
0.6	1.04	0.87	1.05	0.86	1.05	0.85	1.05	0.84
0.8	1.05	0.84	1.06	0.83	1.06	0.82	1.06	0.81
1.0	1.06	0.81	1.07	0.80	1.07	0.79	1.07	0.78
1.5	1.09	0.74	1.09	0.73	1.09	0.72	1.10	0.71
2.0	1.11	0.68	1.11	0.67	1.11	0.66	1.12	0.65
3.0	1.14	0.59	1.14	0.58	1.14	0.57	1.14	0.57
4.0	1.16	0.52	1.16	0.51	1.17	0.50	1.17	0.50

Example 8.3 illustrates how eqns (8.11–8.13) can be applied to determine the plant output for the building illustrated in Example 8.2.

Example 8.3

For the building described in Example 8.2, determine the heating plant output if the building is heated by:

(i)   A forced warm air heating system
(ii) A LTHW panel radiator system
(iii) A high temperature radiant strip system

Assume that the internal design dry resultant temperature is 21 °C and that in all other respects the data are unchanged from that shown in Example 8.2.

Solution

In order quantify F_1 and F_2 the values of the following parameters are determined:

$$\Sigma(A) = 640 \text{ m}^2$$

$$\Sigma(AU) = 381.2 \text{ W/K}$$

$$NV/3 = 800 \text{ W/K.}$$

Therefore

$$\frac{\Sigma(AU)}{\Sigma(A)} = 0.60 \text{ W/m}^2\text{K}$$

and

$$\frac{NV}{3\Sigma(A)} = 1.25 \text{ W/m}^2\text{K}$$

By looking-up Tables 8.8, 8.9 and 8.10 the values of F_1 and F_2 are found to be:

Option	Heating type	F_1	F_2
(i)	Forced warm air heating	0.970	1.010
(ii)	LTHW panel radiator system	1.015	0.965
(iii)	High temperature radiant strip system	1.080	0.755

Therefore, using eqns (8.11) and (8.13):

(i) Forced warm air heating:

$$Q_f = 0.97 \times 381.2 \times (21 - (-1)) = 8134.8 \text{ W}$$

and

$$Q_v = 1.10 \times 0.333 \times 3 \times 800 \times (21 - (-1)) = 19\,360.0 \text{ W}.$$

Therefore

$$Q_p = 8134.8 + 19\,360 = 27\,494.8 \text{ W}$$

Using eqns (8.17–8.19) it is possible to determine the internal air and environmental temperatures, and the mean internal surface temperature of the building.

$$t_{ei} = [0.97 \times (21 - (-1))] + (-1) = 20.3 \text{ °C}$$

$$t_{ai} = [1.01 \times (21 - (-1))] + (-1) = 23.2 \text{ °C}$$

$$t_m = \frac{3}{2} \times 20.3 - \frac{1}{2} \times 23.2 = 18.9 \text{ °C}$$

Similarly the plant output for options (ii) and (iii) can be established using the above methodology. A summary of the results of the winter design calculation for options (i), (ii) and (iii), together with the results from Example 8.2 is presented in the table below.

Option	Heating Type	Q_f (W)	Q_v (W)	Q_p (W)	t_{ai} (°C)	t_m (°C)	t_{ei} (°C)	t_c (°C)
Example 8.2	Non-specific	8386.4	17 600.0	25 986.4	21.0	n.a.	n.a.	n.a.
(i)	Warm air	8134.8	19 360.0	27 494.8	23.2	18.9	20.3	21.1
(ii)	Panel radiators	8512.2	16 983.8	25 496.0	20.2	21.9	21.3	21.1
(iii)	High temp. radiant strip	9057.3	13 287.9	22 345.2	15.6	26.3	22.8	21.0

Comparison of the results from Examples 8.2 and 8.3 indicates that the use of the dry resultant temperature and the F_1 and F_2 coefficients has some effect on the calculated overall plant output. It can be seen that the required output of the high temperature radiant strip system is approximately 5 kW lower than that of the forced warm air system, with the radiator system in the middle. It should also be noticed that the simple method, based on the indoor air temperature, produces a plant output which was similar in magnitude to that required for the LTHW panel radiator system. The margin of error for the less accurate 'air temperature' method, was -5.8% when compared with the warm air system, and $+14.0\%$ when compared with the radiant strip system. It can therefore be concluded that while the simple 'air temperature' method is acceptable for sizing standard wall radiator heating systems, the 'F_1 and F_2' method should be used when designing systems which are either 100% convective or almost 100% radiative.

8.4 Heating energy calculations

Equations 8.5, 8.6, 8.11 and 8.12 demonstrate that building heat loss is directly proportional to the difference in temperature between the internal and external environments. It follows therefore that buildings which experience harsher winters will, not surprisingly, consume more heating energy during the winter months. It is therefore possible to use the degree day method (see Appendix 1 for a detailed explanation of degree days) to predict annual heating costs.

For a building which is continuously heated and which experiences no substantial continuous heat gains, the annual heating energy consumption can be determined by using eqn (8.20) or (8.21).

$$ E = \frac{Q_p}{(t_c - t_{ao})} \times D_{15.5} \times 24 \times \frac{1}{\eta} \qquad (8.20) $$

where E is the energy consumed (kWh), Q_p is the heating plant output (kW), $D_{15.5}$ is the number of standard degree days (i.e. to the base temperature 15.5 °C), and η is the seasonal efficiency of heating system.

Typical values of η for various types of boiler plant are shown in Table 8.11. As an alternative to eqn (8.20), it is possible to use eqn (8.21) to predict heating energy consumption. However, because eqn (8.21) uses internal air temperature and represents a non-specific heating system, the results calculated are likely to be less accurate than those determined using eqn (8.20). Notwithstanding this, eqn (8.21) is reasonably accurate if buildings are well insulated and a predominantly convective heating system is being used.

$$ E = \frac{Q_p}{(t_{ai} - t_{ao})} \times D_{15.5} \times 24 \times \frac{1}{\eta} \qquad (8.21) $$

When designing building heating systems it is standard practice to ignore any internal heat gains in the winter design day calculation. In this way the heating system is sized to meet the 'worst-case' scenario (i.e. when the internal heat gain is not present). In reality, if the heating plant is

Table 8.11   Seasonal heating plant efficiencies [11]

Type of system	Seasonal efficiency (%)
Continuous space heating	
Condenser and conventional boilers with weather compensated system	85
Fully controlled gas or oil fired boiler with radiator system	70
Fully controlled gas or oil fired boiler with radiator system (multiple modular boilers used with sequence controller)	75
Intermittent space heating	
Condenser and conventional boilers with weather compensated system	80
Fully controlled gas or oil fired boiler with radiator system	65
Fully controlled gas or oil fired boiler with radiator system (multiple modular boilers used with sequence controller)	70

over-sized, the heating controls should modulate down the flow water temperature and prevent the building from over-heating. However, when performing energy prediction calculations, it may be important to allow for continuous internal heat gains (e.g. from lighting and equipment) in the degree day calculation. This can be achieved by altering the degree day base temperature from 15.5 °C to an appropriate level. Table 8.12 shows various $D_d/D_{15.5}$ correction factors for various base temperatures.

Internal heat gains can be allowed for in the degree day calculation by determining the temperature rise due to internal gains using eqn (8.22).

$$d = \frac{Q_g}{Q_p} \times (t_c - t_{ao}) \tag{8.22}$$

where d is the average temperature rise which can be maintained by internal heat gains alone (K), and Q_g is the internal heat gain (W).

The new base temperature t_b, can then be determined using eqn (8.23).

$$t_b = t_c - d \tag{8.23}$$

Equations 8.20 and 8.21 only apply to continuously heated buildings. Most buildings are however, intermittently occupied and are not heated continuously. When buildings are intermittently occupied it is necessary to allow for the additional heat energy required to bring the building structure up to temperature. The amount of pre-heating required depends on the thermal capacity of the building.

Table 8.12 $D_d/D_{15.5}$ ratios for various base temperatures [12]

Base temperature (°C)	$D_d/D_{15.5}$
10	0.33
12	0.57
14	0.82
15	0.94
15.5	1.00
16	1.06
17	1.18
18	1.30

Heavy structures require long pre-heat periods and, once heated, retain heat well, while lightweight structures tend to heat-up and cool quickly. It is therefore impossible to fully consider the impact of intermittent occupation on energy consumption without considering the thermal capacity of the building. The CIBSE classification of structures by thermal inertia is presented in Table 8.13.

To allow for intermittent heating when using the degree day method it is necessary to introduce correction factors for:

- The length of the working week.
- The length of the working day.
- The response of the building and plant.

Tables 8.14, 8.15 and 8.16 set out values for these correction factors. Example 8.4 illustrates how the annual heating costs for a building can be calculated.

Example 8.4

A three-storey office building, with a total floor area of 2400 m^3, is occupied for 5 days per week and for 8 hours per day. The design day plant output (i.e. heat loss) is calculated to be 190.0 kW when the external temperature is -3.0 °C and the dry resultant temperature is 21 °C. The building is heated by a series of gas-fired modular boilers connected to a responsive warm air heating system with a seasonal efficiency of 70%.

Given that the building is located in a region which experiences 2354 degree days per year and that the cost of natural gas is 1.5 p/kWh, determine the annual heating fuel cost:

(i) Ignoring any internal heat gains (i.e. assuming a base temperature of 15.5 °C)
(ii) Allowing for an internal heat gain (from lights and equipment) of 20 W/m^2.

Solution

(i) *Ignoring any internal heat gains*: From Table 8.14, the correction factor for length of working week is 0.85. From Table 8.15, the correction factor for length of working day is 1.00. From Table 8.16, the correction factor for response of building and plant is 0.85.

Table 8.13   CIBSE classification of structures by thermal inertia [12]

Weight	Building description
Very heavy	Multi-storey buildings with masonry or concrete curtain walling and sub-divided within by solid partitions.
Heavy	Buildings with large window areas but appreciable areas of solid partitions and floors.
Medium	Single-storey buildings of masonry or concrete, sub-divided within by solid partitions.
Light	Single-storey buildings of a factory type, with little or no solid partitions.

Table 8.14   Correction factor for length of working week [13]

Occupied days per week	Lightweight building	Heavyweight building
7 days	1.0	1.0
5 days	0.75	0.85

Table 8.15　Correction factor for length of working day. Applies to Intermittent use only [13]

Occupied period	Lightweight building	Heavyweight building
4 hours	0.68	0.96
8 hours	1.00	1.00
12 hours	1.25	1.02
16 hours	1.40	1.03

Table 8.16　Correction factor for the response of building and plant [13]

Type of heating	Lightweight	Medium weight	Heavyweight
Continuous	1.0	1.0	1.0
Intermittent – responsive plant	0.55	0.70	0.85
Intermittent – plant with a long time lag	0.70	0.85	0.95

Therefore:

$$\text{Annual heating energy consumption} = \frac{190}{(21 - (-3))} \times 2354 \times 24 \times \frac{(0.85 \times 1.00 \times 0.85)}{0.7}$$

$$= 461\ 636.21 \text{ kWh}$$

and

$$\text{Annual energy cost} = \frac{461636.21 \times 1.5}{100} = £\ 6924.54$$

(ii) *Allowing for an internal heat gain of* 20 W/m^2:

$$\text{Total heat gain} = \frac{2400 \times 20}{1000} = 48.0 \text{ kW}$$

Using eqn (8.22), the temperature rise due to heat gains is:

$$d = \frac{48}{190} \times (21-(-3)) = 6.06\ °C$$

Therefore, the new base temperature is:

$$t_b = 21 - 6.06 = 14.94\ °C$$

From Table 8.12 this corresponds to a $D_d/D_{15.5}$ value of 0.932. Therefore

$$\text{Annual heating energy consumption (allowing for heat gains)} = 461\ 636.21 \times 0.932$$
$$= 430\ 244.95 \text{ kWh}$$

Therefore

$$\text{Annual energy cost} = \frac{430\,244.95 \times 1.5}{100} = £\,6453.67$$

As well as predicting heating energy costs, it is also possible to use the degree day method to evaluate proposed energy saving measures. Example 8.5 illustrates how this can be achieved.

Example 8.5

An office building has a roof which has a U value of 1.1 W/m²K. It is proposed that additional insulation be installed in the roof to bring its U value down to 0.25 W/m²K.

The office building is located in a region which experiences an annual total of 2350 degree days. Assuming that the efficiency of the building heating system is 70%, the cost of fuel is 1.5 p/kWh and the capital cost of installing the roof insulation is £2.00 per m², determine the payback on the investment.

Solution

$$\text{Annual energy saving} = (1.1 - 0.25) \times \frac{2350}{100} \times \frac{24}{0.7}$$
$$= 68.486 \text{ kWh/m}^2.$$

Therefore

$$\text{Annual energy cost saving} = \frac{68.486 \times 1.5}{100} = £1.03 \text{ per m}^2$$

$$\text{Payback period} = \frac{2.00}{1.03} = 1.94 \text{ years}$$

8.5 Intermittent heating

The degree day calculations in Section 8.4, indicate that the intermittent use of heating plant results in higher energy consumption compared with continuous heating. This is primarily because intermittently occupied buildings, such as office buildings, require the structure to be warmed up after it has been allowed to cool over-night and at weekends. Additional heat energy is therefore required to 'pre-heat' the building in the mornings, so that its fabric is brought up to a temperature which will be comfortable for the occupants. Buildings with a high thermal mass require greater pre-heating than buildings with a low thermal mass. However, once heated to the required temperature, heavyweight structures retain their heat for much longer than lightweight ones.

Table 8.17 Recommended plant ratios for intermittent heating [10]

Plant ratio (F3)	Lightweight building pre-heat time (hours)	Heavyweight building pre-heat time (hours)
1.0	Continuous	Continuous
1.2	6	Very long
1.5	3	7
2.0	1	4
2.5	0	2
3.0	0	1

The pre-heating of a building structure is achieved by running the heating system at full capacity for a pre-heat period prior to the building's occupation. The heavier the building structure, the longer the pre-heat period. The pre-heat period can be reduced in length by over-sizing the boiler plant and, to some extent, the heat emitters. It is generally considered more energy efficient to increase the plant margin (i.e. over-size the boiler plant) in order to reduce the pre-heat period [10]. Table 8.17 gives recommended plant ratios for intermittent heating [10], which should be used in conjunction with eqn (8.24) to determine the intermittent peak heating load. It should be noted that the pre-heat times in the table assume the use of plant with a short response time, such as a warm air heating system. For slow response heating systems such as under-floor heating the pre-heating period will be longer.

$$Q_{pb} = F_3 \times Q_p \tag{8.24}$$

where Q_{pb} is the intermittent peak heating output (W or kW), and F_3 is the plant ratio (i.e. Max. heat output / Design day heat output).

When over-sizing plant it is important to consider both boilers and heat emitters. While increased boiler capacity may reduce the pre-heat time, it can result in poor part-load performance and low seasonal boiler efficiency. It should be remembered that for most of the heating season, the external air temperature will be well above the winter design condition, so for much of the year the boilers will have plenty of excess capacity. It is therefore wise to consider the use of modular boilers.

While increased boiler capacity may be advisable, it is not always necessary to increase individual heat emitters. This is because the following alternative strategies can be employed:

- In buildings which are unoccupied, natural and mechanical ventilation rates can be reduced during the nighttime. Reduced ventilation rates will occur naturally during the nighttime because doors and windows usually remain closed. With mechanical ventilation systems it is also possible to reduce the ventilation load by fully recirculating the air (i.e. reducing the outside air component of the supply air to 0%) during the pre-heat period.
- It is possible to elevate the water supply temperature to the heat emitters during the pre-heat period.

Although, a modest over-sizing of heat emitters is recommended by the CIBSE, there is no strong economic case for considerable over-sizing (i.e. in excess of 25%) of emitter surfaces [10].

8.6 Radiant heat

The importance of radiant heat transfer in buildings is often misunderstood with the result that potential 'radiant' energy saving measures are often ignored. It is therefore worth investigating some of the 'radiant' energy saving techniques that exist.

8.6.1 Radiant heating

Equation 8.4 shows that the comfort of building occupants is as dependent on the mean radiant temperature as it is on air temperature. This fact can be used to great advantage in applications where a building is poorly insulated and in which ventilation rates are high, such as in old factories or workshops. In such applications, it is often prohibitively expensive to heat up large volumes of air, which are then quickly lost to the outside. It is much better to use some form of radiant heating to warm up the occupants. By using a high temperature radiant heat source it is possible to create a heat balance which enables the occupants to feel comfortable whilst still maintaining the air and fabric at a low temperature.

Radiant heating is particularly well suited to applications in which occupancy is very intermittent and in which the occupants are located in relatively fixed positions. A church building is a classic example of such an application. Such a building is occupied for a relatively short period in every week. Because radiant heating systems react very quickly and warm the occupants rather than the air, they can achieve a good comfort level without any pre-heating of the building. For this and the other reasons mentioned, radiant heating systems are generally considered to incur lower capital costs and lower operating costs than other comparable systems [13].

In order to achieve high levels of radiant heat transfer it is necessary to have emitters which are at a high temperature, well above 100 °C. For safety and comfort reasons, these heat emitting panels must be placed at high level, well out of the reach of any of the occupants.

8.6.2 Low emissivity glazing

Glazing is often viewed as a thermal 'weak-link', because heat is easily conducted through glass from the inside to the outside. This perception is broadly, but not wholly, true. It is often forgotten that much of the heat which is lost through windows occurs because they are, in effect, large flat high emissivity surfaces which are cold in relation to the other surfaces in a room. All the other surfaces in the room, especially heated surfaces, such as radiators and warm bodies, emit long-wave radiation which is readily absorbed by the cool glass. In this way much heat is lost from buildings. It is possible to minimize this problem by installing low emissivity glazing which reduces the absorption of long-wave radiation. The low emissivity effect is achieved during the manufacturing process by applying a microscopically thin (i.e. 0.3–0.4 μm thick) coating of tin oxide doped with fluorine atoms [14] to the cavity-facing surface of the inner pane of a double-glazed unit. This significantly reduces the radiative heat loss to the cavity and thus reduces the U value of a typical double glazed unit from approximately 3 W/m^2K to 1.8 W/m^2K.

8.7 Under floor and wall heating

It is possible to heat buildings efficiently by using systems which utilize very low water temperatures. Two effective low temperature systems which can be used are:

- Under floor heating, which utilizes flow water temperatures in the range 35 °C–50 °C.
- Wall heating, which utilizes flow water temperatures in the range 30 °C–40 °C.

The low water temperatures involved in these systems enable alternative heat sources to be utilized, such as ground source heat pumps, or even solar energy. Because both wall and under floor heating systems involve large heated surfaces, the room mean radiant temperature is increased. This makes it possible to reduce the air temperature within the room space without altering the *dry resultant temperature*, enabling energy to be saved whilst still maintaining a comfortable environment.

The maximum permissible surface temperature governs the maximum water flow temperature which can be used in both wall and under-floor heating installations. In the case of under-floor heating, occupants feel uncomfortable if the floor surface temperature is above 29 °C. For walls the maximum safe surface temperature is about 43 °C, since the disassociation temperature for plaster is approximately 45 °C.

Under-floor heating systems are best suited to tall spaces, where the use of a conventional warm air or radiator system may lead to stratification of the air (i.e. the warm air becoming trapped at the top of the room space). The use of under-floor heating overcomes this problem and ensures that the air is warmest at ground level, where the occupants are likely to be located.

Under-floor heating systems often consist of a continuous cross-linked polyethylene, or polypropylene flexible pipe loop embedded in a floor screed, on top of a structural floor [15]. The screed is usually isolated from the structural floor by rigid insulation slabs, which reduce the heat transfer through the slab and help to maintain the screed temperature. The nature of the screed used in under-floor heating systems is of particular importance, since the screed acts as a thermal resistance to the heat transfer from the pipe to the room space, and also provides the system with thermal inertia. Screeds can be either cementitious in nature and approximately 75 mm thick, or an anhydrite flowing screed which enables the thickness to be reduced to approximately 50 mm. The thermal storage capacity and inertia of a floor screed depend on a number of parameters such as the specific heat capacity, screed thickness, screed density, thermal conductivity, pipe spacing, and the water flow and return temperatures. It is therefore difficult to predict exactly how any given floor will perform in practice. However, for a standard 65 mm thick concrete screed floor, with a flow water temperature of 60 °C, it has been estimated that the floor screed will take approximately 3 hours to charge, and 3 hours to discharge. At lower water temperatures the charging process is considerably extended. The long hysteresis effect of under-floor heating has the disadvantage of being slow to respond and makes it difficult to adjust to sudden changes in the internal environment. Under-floor heating systems are therefore best suited to applications in which occupancy is continuous or well-defined and predictable.

Wall heating operates in a similar manner to under floor heating, with the exception that the pipe coils are generally not located in a material which has a high thermal mass, such as a floor screed. In a typical wall heating system cross-linked polyethylene pipes are located on the air cavity side of a dry lined plastered wall. The system can be used either with a wet plastered board, or alternatively the pipes can be mechanically bonded to plasterboard using notched battens to make a rigid unit which can then be fixed to the wall. A flexible insulation quilt, such as rockwool, should be placed between the pipes and the structural wall, so that conduction to the building structure is minimized. It is advisable where possible to install the wall heating on internal walls, so that heat losses are minimized.

8.8 Pipework insulation

Considerable amounts of heat energy can be lost through uninsulated or poorly insulated pipework. It is therefore important to ensure that hot water and steam pipework is properly

Table 8.18   Thermal conductivities of insulating materials [16]

Material	Density (kg/m³)	Thermal conductivity (W/mK)		
		50 °C	100 °C	300 °C
Calcium silicate	210	0.055	0.058	0.083
Expanded nitrile rubber	65–90	0.039	–	–
Mineral wool (glass)	16	0.047	0.065	–
Mineral wool (rock)	100	0.037	0.043	0.088
Magnesia	190	0.055	0.058	0.082
Polyisocyanurate foam	50	0.023	0.026	–

insulated. A range of insulating materials is available and these can be either inorganic, based on crystalline or amorphous silicon, aluminium or calcium, or organic, based on hydrocarbon polymers in the form of thermosetting/thermoplastic resins or rubbers [16]. They can be either flexible or rigid, both types being available in pre-formed pipe sections. Table 8.18 lists some of the common types of insulation along with some of their thermal properties.

8.8.1 Pipework heat loss

In Section 8.3.1 heat loss through flat surfaces such as walls and roofs is discussed. The geometry of pipework is different from that of flat surfaces, requiring an alternative approach when calculating the heat loss from sections of pipework. The heat transfer through the wall of a pipe can be calculated using eqn (8.25).

$$Q = \frac{2\pi\lambda(t_1 - t_2)}{\ln(r_2/r_1)} \quad \text{(W per metre length)} \tag{8.25}$$

where λ is the thermal conductivity of pipe wall (W/mK), r_1 is the internal radius of pipe (m), and r_2 is the external radius of pipe (m).

The thermal resistance of the pipe wall (per unit length of pipe) can be determined by:

$$R = \frac{\ln(r_2/r_1)}{2\pi\lambda} \quad \text{(mK/W)} \tag{8.26}$$

As hot fluid flows along a pipe, heat is transferred to the pipe wall. The rate at which this heat is transferred depends on the thermal resistance of a thin stationary layer of fluid on the pipe wall surface. The heat transfer rate across this internal surface boundary layer can be expressed as:

$$Q = h \times A \times \Delta t \tag{8.27}$$

where h is the surface heat transfer coefficient (W/m^{2}K), A is the surface area (m^2), and Δt is the temperature difference between the surface and the bulk fluid (°C).

Equation (8.27) can also be applied to the heat transfer across the external surface of a pipe. Consequently, the internal and external surface resistance per unit length of a pipe can be expressed as:

$$R_{\text{so or si}} = \frac{1}{h.A} \quad \text{(mK/W)} \tag{8.28}$$

The overall resistance per unit length of a typical insulated pipe can therefore be represented by:

$$R_t = R_{si} + R_w + R_{ins} + R_{so} \tag{8.29}$$

where R_t is the total thermal resistance of pipework per unit length (mK/W), R_w is the thermal resistance of pipe wall per unit length (mK/W), R_{ins} is the thermal resistance of insulation per unit length (mK/W), R_{si} and R_{so} are the internal and external surface thermal resistances of insulation per unit length (mK/W).

Once the overall resistance of the pipework is determined, the total heat loss per metre run can be calculated by dividing the temperature difference between the fluid and ambient air by the total resistance.

$$Q = \frac{\Delta t}{R_t} \quad \text{(W/m)} \tag{8.30}$$

Example 8.5

A pipe carries wet steam at 200 °C through a building which has an ambient air temperature of 20 °C. The pipe has an internal diameter of 53.5 mm, a wall thickness of 3.7 mm, and is insulated to a thickness of 25 mm. The thermal conductivity of the pipe material is 46 W/m K and the thermal conductivity of the insulating material is 0.033 W/mK. Assuming that the inside and outside surface heat transfer coefficients are 10 000 W/m² K and 10 W/m² K respectively, determine:

(i) The heat transfer per metre length of pipe;
(ii) The temperature of the outside surface of the insulation;
(iii) The heat loss from an uninsulated pipe, assuming that the external heat transfer coefficient remains unchanged.

Solution

(i) *The total pipework resistance per metre length is:*

$$R_t = R_{si} + R_w + R_{ins} + R_{so}$$

and

$$R_{si} = \frac{1}{10000 \times \pi \times 0.0535} = 0.00059 \text{ mK/W}$$

$$R_{so} = \frac{1}{10 \times \pi \times 0.1109} = 0.287 \text{ mK/W}$$

$$R_w = \frac{\ln (30.45/26.75)}{2 \times \pi \times 46} = 0.00045 \text{ mK/W}$$

$$R_{ins} = \frac{\ln (55.45/30.45)}{2 \times \pi \times 0.033} = 2.891 \text{ mK/W}$$

From the above calculation it can be seen that the resistance of the pipe wall is negligible compared with that of the insulation and the outside surface resistance. In addition the heat transfer coefficient for the internal surface is very high and hence the internal surface resistance is negligible. Therefore, the total thermal resistance can be assumed to be:

$$R_t = R_{ins} + R_{so}$$

$$= 3.178 \text{ mK/W}$$

Therefore

$$Q = \frac{(200 - 20)}{3.178} = 56.64 \text{ W/m}$$

(ii) *The surface temperature can be found by applying the following equation:*

Temperature of the outside of the insulation $= t_a + (R_{so} \times Q)$
$$= 20 + (0.287 \times 56.64) = 36.3 \text{ °C}$$

where t_a is the room air temperature (°C).

(iii) *For the uninsulated pipe*:

$$R_t = R_{si} + R_w + R_{so}$$

$$= 0.00059 + 0.00045 + 0.287 = 0.288 \text{ mK/W}$$

therefore

$$Q = \frac{(200 - 20)}{0.288} = 625 \text{ W/m}$$

As well as illustrating the mechanics of a pipework heat loss calculation, Example 8.5 illustrates the great benefit to be derived from insulation, since the 25 mm insulation layer reduced the heat loss from 625 to 56.6 W/m.

8.8.2 Economics of pipework insulation

It is well known that one of the simplest and most cost effective ways of preventing energy wastage is to insulate pipework runs. Nevertheless it is not easy to determine to what extent pipework should be insulated. The capital cost of insulation increases with its thickness and the financial saving must be off-set against the capital cost. The economic thickness of insulation should therefore be governed by the pay-back period that is required on the investment, as illustrated in Example 8.6.

Example 8.6

A steel pipe carries high pressure hot water at 120 °C around a factory building. The owners of the factory propose to insulate the pipe using rock-wool. Given the data below, determine the optimum thickness of the insulation and the simple pay-back period for that thickness.

Data:

Pipe outside diameter = 76.6 mm
Heat transfer coefficient for outside surface insulation = 10 W/m^{2}K
Thermal conductivity of insulation = 0.037 W/m K
Water temperature = 120 °C
Temperature of air in factory = 15 °C
Boiler efficiency = 70%
Unit price of gas = 1.52 p/kWh
Boiler operates for 2500 hours per annum

Insulation costs

Thickness of insulation (mm)	[20]	[25]	[32]	[38]	[50]	[60]	[75]
Cost per metre length (£/m)	5.00	5.58	6.64	8.01	10.57	13.44	16.68

Assume that the write-off period for the insulation is 5 years.

Solution

Assuming that the thermal resistances of the pipe wall and of the inside surface of the pipe are both negligible, let

$$x = \text{thickness of insulation (mm)}$$

Now

$$R_{ins} = \frac{\ln(r_2/r_1)}{2\pi\lambda}$$

$$= \frac{\ln((38.3 + x)/38.3)}{2 \times \pi \times 0.037}$$

and

$$R_{so} = \frac{1}{h.A}$$

$$= \frac{1}{10 \times \pi \times ((76.6 + 2x) \times 10^{-3})}$$

$$R_t = R_{ins} + R_{so}$$

and

$$Q = \frac{\Delta t}{R_t}$$

therefore

$$Q = \frac{(120 - 15)}{R_{ins} + R_{so}}$$

and

$$\text{Annual operating cost} = \frac{Q \times 2500 \times 1.52}{0.7 \times 100}$$

and

$$\text{Total annual cost} = \text{annual fuel cost} + \frac{\text{capital cost}}{\text{write-off period}}$$

The following table can be produced from the above equations:

Insulation thickness (mm)	Insulation resistance (mK/W)	External resistance (mK/W)	Heat loss (W/m)	Annual fuel cost (£/m)	Total annual cost (£/m)
0	0.00	0.42	252.68	13.72	13.72
20	1.81	0.27	50.47	2.74	3.74
25	2.16	0.25	43.52	2.36	3.48
32	2.61	0.23	36.99	2.01	3.34
38	2.96	0.21	33.09	1.80	3.40
50	3.59	0.18	27.83	1.51	3.62
60	4.05	0.16	24.90	1.35	4.04
75	4.67	0.14	21.85	1.19	4.52

It can be seen that the most economic thickness of insulation is 32 mm, since this has the lowest annual cost.

The pay back period for the 32 mm insulation is:

$$\text{Payback period} = \frac{6.62}{(13.72 - 3.34)} = 0.638 \text{ years}$$

8.9 Boilers

Most heating systems, although not all, employ boilers to produce hot water or steam. Boiler efficiency therefore has an important influence on heating-related energy costs. The cost savings that can be achieved by improving overall boiler efficiency can be substantial. Essentially a boiler is a device in which a fossil fuel is burnt and the heat produced is transferred to water. The more effective this heat transfer process, the more efficient the boiler. It is therefore important to maximize the heat transfer to the water and minimize boiler heat losses. Heat can be lost from boilers by a variety of methods, including flue gas losses, radiation losses and, in the case of steam boilers, blow-down losses. Although all these various losses have a significant effect on boiler energy consumption, the major reason for poor boiler performance occurs at the design stage, where the capacity of boilers is usually oversized and inappropriate boilers are often selected.

Boiler plant which is oversized will operate at part-load for most of the time, resulting in low seasonal efficiency and high operating costs. It has been estimated that a 15% increase in energy consumption can occur if a conventional boiler plant is oversized by 150% [17]. Boiler plant should be considered oversized if under the winter design condition the boilers are able to maintain an internal air temperature well above the design temperature (e.g. 21 °C). Evidence of over-sizing can be manifested in a number of ways: fuel bills may be higher than expected; excessive cycling of boiler plant may be experienced; and in installations equipped with modular boilers, a large proportion of the boilers may never be used.

Given that for much of the heating season external air temperatures are usually well above the winter design condition, it is important that any boiler installation be designed so that it operates efficiently at part load. For most types of conventional boiler, efficiency falls dramatically below about 30% of rated capacity [18]. Large boilers are therefore at a disadvantage since for most of the time they will be operating well below their rated capacity. One simple way of overcoming this problem is to install a large number of small modular boilers with a sequence controller in preference to a few large boilers. This ensures that under part-load conditions, boilers which are always operating near their maximum efficiency will provide the bulk of the heating. With such a multi-boiler plant installation, it is wise to install a boiler sequence control system. This is a fully automatic microprocessor-controlled system which monitors and sequences on/off operations of boiler plant according to the demand for heat. This avoids running too many boilers on part-load and minimizes the number of boilers in operation at any one time.

Another technique which can be employed to ensure good part-load efficiencies is to use boilers with modulating burners. These burners modulate the fuel and air to provide a variable output from 20–30% to 100%. With large modulating boilers it is possible to make substantial energy savings by installing variable speed drives on the combustion air fans. Variable speed drives reproduce the operating characteristics of fixed speed combustion air fans and adjustable dampers, whilst reducing the average electrical demand of the fan motor by approximately 60% [19].

8.9.1 Flue gas losses

All boilers require a minimum amount of air to ensure that complete combustion of the fuel takes place and that no carbon monoxide is produced. Yet, if too much air is added then heat is wasted in warming up the excess air, which then escapes through the flue. The amount of combustion air should therefore be limited to that necessary to ensure complete combustion of the fuel. In practice some excess air, around 15–25% for oil fired boilers [19], and 15–30% for gas fired boilers [20] is needed. The actual amount required to give the optimum boiler efficiency depends on the fuel used and the type of boiler and burner. If the air flow rate to a boiler is too low, then a proportion of the fuel will remain unburnt and running costs will increase. In the case of oil-fired plant, incomplete combustion will produce smoke which will be visible. For coal fire plant, incomplete combustion results in unburnt carbon in the ash. It is therefore essential to maintain the correct fuel-to-air ratio at all times. With modern microprocessor controlled burners, which are fitted to fuel valves and air dampers, it is possible to automatically select and maintain specific fuel-to-air ratios for a variety of fuels. These controllers continually monitor the level of oxygen in the flue gases, and alter the combustion air supply in order to maintain optimum conditions.

Flue gas losses are by far the greatest heat losses which occur from boilers. The flue gases contain considerable sensible heat and also latent heat which is 'bound up' in water vapour. It is possible to determine the amount of heat which is being lost through the flue, by monitoring the presence of O_2 or CO_2 in the flue gases. If there is little excess air in the combustion gases, then the percentage of CO_2 will be high and the percentage of O_2 low. Conversely, if a large amount of excess air is present, the relationship will be reversed. In a typical gas-fired shell and tube boiler the flue gases should contain about 9–10% CO_2 and 3–5% O_2 [20], while for an oil fired boiler the CO_2 content of the flue gases should be in the region of 13–14% [19]. Typical CO_2 and O_2 flue gas contents for efficient boiler operation are presented in Table 8.19.

With large boiler plant it is possible to increase boiler efficiency by pre-heating the combustion air. It has been estimated that the thermal efficiency of a boiler can be increased by approximately 1% if the temperature of the combustion air is raised by 20 °C [19, 20]. Any one of the following sources of heat can be utilized to pre-heat the boiler combustion air:

- Waste heat from the flue gases.
- Drawing high temperature air from the top of the boiler room.
- Recovering waste heat by drawing air over or through the boiler casing.

8.9.2 Other heat losses

With shell and tube boilers it is possible for the 'smoke' tubes to become fouled by soot and other deposits, resulting in a reduction in the amount of heat which is transferred from the hot flue gases to the water. This increases the temperature of the flue gases and results in greater flue gas losses. Boiler smoke tubes should therefore be cleaned regularly to minimize the flue gas temperature rise, since it has been estimated that a rise of 17 °C in the flue gas temperature causes

Table 8.19   Typical CO_2 and O_2 contents by volume expected in flue gas (dry basis) [18]

| Fuel | Min. fire | | Full fire | | |
	CO_2 (%)	O_2 (%)	CO_2 (%)	O_2 (%)	CO (ppm)
Coal	11.0	8.5	14.0	5.0	2–500
Fuel oils	11.5	5.5	13.5	3.0	–
Butane	9.4	7.0	12.0	3.0	2–400
Propane	9.2	7.0	12.0	3.0	2–400
Natural gas	8.0	7.0	10.0	3.3	2–400

a decrease in efficiency of approximately 1% [19]. Boiler efficiency can also significantly be reduced by a build-up of scale on the water side of the smoke tubes. Water treatment should therefore be employed in order to prevent scale formation.

Heat can be lost through the surface casing of boilers. This is generally referred to as *radiation loss*, although it includes heat which is lost by convection. With modern boilers radiation losses are usually not greater than 1% of the maximum rating. On older boilers this figure may be as high as 10% where the insulation is in poor condition [19, 20].

8.9.3 Boiler blow-down

With steam boilers it is necessary to eject a small proportion of the water regularly in order to remove sludge and to maintain acceptable levels of total dissolved solids. This process is called *blow-down* and it prevents scaling up of the tubes on the water side. Although necessary, blow-down represents a considerable energy loss and the level of blow-down should be kept to a minimum while still maintaining the recommended level of dissolved solids.

It is possible to recover waste heat from the blow-down process by collecting the flash steam which forms as the pressure falls through the blow-down valve. Because the condensate produced by the blow-down process is both hot and pure, with no dissolved solids, it can be added directly to the make-up water for the boiler, thus reducing energy consumption.

8.9.4 Condensing boilers

Boiler flue gases are often in excess of 200 °C and as such are a useful source of waste heat recovery. Heat exchangers can be installed in flues to recover both sensible and latent heat from the hot products of combustion. However, because of the corrosion problems associated with sulphur bearing fuels, such as fuel oil, flue gas heat recovery is generally only practised on gas-fired boilers. Gas-fired boilers which incorporate integral flue gas heat exchangers are known as *condensing boilers*. If used correctly, condensing boilers can achieve operating efficiencies in excess of 90% [18, 21].

With a condensing boiler it is desirable to operate the system so that the return water temperature is as low as possible. This ensures that condensation of the flue gases occurs and that maximum heat recovery from flue gases is achieved. If a condensing boiler is used in conjunction with a weather compensating controller, the boiler will move into condensing mode during the milder part of the season, when return water temperatures are at their lowest. In this way, high efficiency is maintained under part-load conditions. In multiple boiler installations it is usually cost effective to install only one condensing boiler. This should always be the lead boiler, since it exhibits the highest efficiencies. This ensures that good energy utilization will occur under part-load conditions.

References

1. Keating, W. R. and the Eurowinter Group (1997). Cold exposure and winter mortality from ischaemic heart disease, cerebrovascular disease, respiratory disease, and all causes in warm and cold regions of Europe. *The Lancet*, 349, 1341–6.
2. Forecasting the nation's health (2000). Met Office Report, November.
3. Wilkinson, P. (2000). Social and environmental determinants of excess winter deaths in England, 1986–1996. Report for the Joseph Rowntree Foundation, July.
4. Edholm, O. G. (1978). *Man – hot and cold (Studies in Biology No. 97)* (Chapter 1). Arnold.

5. CIBSE Guide A1, Environmental criteria for design (1986).
6. Smith, B. J., Phillips, G. M. and Sweeney, M. (1987). *Environmental science* (Chapter 2). Longman Scientific and Technical.
7. CIBSE Guide A2, Weather and solar data (1982).
8. CIBSE Guide A3, Thermal properties of building structures (1980).
9. Smith, B. J., Phillips, G. M. and Sweeney, M. (1987). *Environmental science* (Chapter 1). Longman Scientific and Technical.
10. CIBSE Guide A9, Estimation of plant capacity (1983).
11. Moss, K. J. (1997). *Energy Management and Operating Costs in Buildings*, (Chapter 1). E & FN Spon.
12. CIBSE Guide B18, Owning and operating costs (1986).
13. Economic use of fired space heaters for industry and commerce, Fuel Efficiency Booklet 3 (1993). Department of the Environment.
14. Enhancing the built environment. Pilkington Plc. website.
15. CIBSE Guide B1, Heating (1986).
16. The economic thickness of insulation for hot pipes, Fuel Efficiency Booklet 8 (1994). Department of the Environment.
17. Hendrick, R. et al. (1992). Furnace sizing criteria for energy-efficient set back strategies. *ASHRAE Transactions*, **98**(Part 1), 1239–46.
18. Energy audits and surveys (1991). CIBSE Applications Manual AM5.
19. Economic use of oil-fired boiler plant, Fuel Efficiency Booklet 14 (1993). Department of the Environment.
20. Economic use of gas-fired boiler plant, Fuel Efficiency Booklet 15 (1994). Department of the Environment.
21. Condensing Boilers (1989). CIBSE Applications Manual AM3.

Bibliography

CIBSE Guide A1, Environmental criteria for design (1986).

CIBSE Guide A2, Weather and solar data (1982).

CIBSE Guide A3, Thermal properties of building structures (1980).

CIBSE Guide A9, Estimation of plant capacity (1983).

CIBSE Guide B1, Heating (1986).

CIBSE Guide B18, Owning and operating costs (1986).

Degree days, Fuel Efficiency Booklet 7 (1995). Department of the Environment.

Eastop, T. D. and Croft, D. R. (1990). *Energy efficiency for engineers and technologists* (Chapters 4 and 5). Longman Scientific & Technical.

Eastop, T. D. and Watson, W. E. (1992). *Mechanical services for buildings* (Chapter 3). Longman Scientific & Technical.

Edholm, O. G. (1978). *Man – hot and cold (Studies in Biology No. 97)* (Chapter 1). Arnold.

Economic use of fired space heaters for industry and commerce. Fuel Efficiency Booklet 3 (1993). Department of the Environment.

Economic use of oil-fired boiler plant, Fuel Efficiency Booklet 14 (1993). Department of the Environment.

Economic use of gas-fired boiler plant, Fuel Efficiency Booklet 15 (1994). Department of the Environment.

Levermore, G. J. (1992). *Building energy management systems* (Chapter 10). E & FN Spon.

Moss, K. J. (1997). *Energy Management and Operating Costs in Buildings* (Chapters 1 and 3). E & FN Spon.

Stoecker, W. F. and Jones, J. W. (1982). *Refrigeration and air conditioning* (Chapters 2 and 4). McGraw-Hill International Editions.

The economic thickness of insulation for hot pipes, Fuel Efficiency Booklet 8 (1994). Department of the Environment.

9

Waste heat recovery

In many applications there is the potential for recovering heat energy that would otherwise go to waste. This chapter describes various waste heat recovery technologies and examines the theoretical principles behind each.

9.1 Introduction

In many applications it is possible to greatly reduce energy costs by employing some form of waste heat recovery device. However, before investing in such technology it is important to first consider some generic issues:

- Is there a suitable waste heat source? If the answer to this question is 'yes', it is important to establish that the source is capable of supplying a sufficient 'quantity' of heat, and that the heat is of a good enough 'quality' to promote good heat transfer.
- Is there a market or use for the recovered waste heat? It is important to have a use for any waste heat which may be recovered. In many applications there may be no demand for the heat that is available, with the result that a large quantity of heat energy is dumped. In other situations there may be a long time lag between waste heat production and the demand for heat. Waste heat therefore cannot be utilized unless some form of thermal storage is adopted.
- Will the insertion of a heat recovery device actually save primary energy or reduce energy costs? Often the insertion of a heat exchanger increases the resistance of the fluid streams, resulting in fan or pump energy consumption rising. Heat energy is therefore replaced by electrical energy, which may be produced at an efficiency of less than 35%.
- Will any investment in heat recovery technology be economic? Heat recovery devices can be expensive to install. It is, therefore, essential that the economic payback period be determined prior to any investment being undertaken.

Although the questions above may appear obvious, it is not uncommon to find cases where poor planning and analysis at the design stage has resulted in an installation where the impact of the heat recovery device is either minimal, or is even increasing energy costs. Consider the case of a heat exchanger installed in a warm exhaust air stream from a building. The insertion of the device causes the resistance of the air streams to rise, resulting in greater fan energy consumption. If the unit price of electricity is four times that of gas, then in order to just break even, the heat exchanger must recover four times the increase in electrical energy consumption, arising from the increased system resistance. Also, there may well be long periods when the external air temperature is such that little or no heat can be recovered. If however, the fans run continuously then the increased electrical energy is being expended for little or no return. Given this, it is not surprising that many so-called 'energy recovery' systems, while appearing to save energy, are in fact increasing both primary energy consumption and energy costs.

If a strategic decision is made to invest in some form of heat recovery device, then the next logical step is to select the most appropriate system. There are a wide variety of heat recovery technologies, which can be divided into the following broad categories:

- *Recuperative heat exchanges*: where the two fluids involved in the heat exchange, are separated at all times by a solid barrier.
- *Run-around coils*: where an independent circulating fluid is used to transport heat between the hot and cold streams.
- *Regenerative heat exchanges*: where hot and cold fluids pass alternately across a matrix of material.
- *Heat pumps*: where a vapour compression cycle is used to transfer heat between the hot and cold streams.

In addition to these, there are a few lesser used technologies, such as heat pipes, which are not covered in this text.

9.2 Recuperative heat exchangers

In a recuperative heat exchanger the two fluids involved in the heat transfer are separated at all times by a solid barrier. This means that the mechanisms which control the heat transfer are convection and conduction. The thermal resistance of a heat exchanger can therefore be expressed as follows:

$$R = \frac{1}{U} = \frac{1}{h_i} + R_w + \frac{1}{h_o} + F_i + F_O \tag{9.1}$$

where R is the thermal resistance of the heat exchanger (m²K/W), R_w is the thermal resistance of the separating wall (m²K/W), h_i and h_o are the heat transfer coefficient of internal and external surfaces (W/m²K), F_i and F_o are the internal and external fouling factors, and U is the overall heat transfer coefficient (i.e. U value) (W/m²K).

In short this can be written as:

$$\frac{1}{U_{dirty}} = \frac{1}{U_{clean}} + \text{Fouling factors} \tag{9.2}$$

In practice heat exchangers are often over-sized so that even when fouled their performance still meets design requirements. The degree of over-sizing is achieved by incorporating fouling factors into the sizing calculation.

Recuperative heat exchangers are the most common type of equipment used for waste heat recovery. They can only be used in applications where the hot and cold streams can be brought into close proximity with each other. Although the precise form of a heat exchanger may change with its particular application, there are three forms which are widely used.

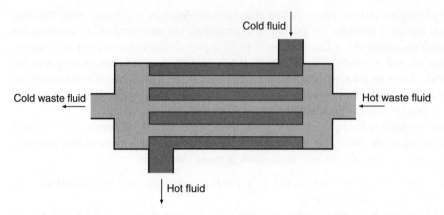

Cold fluid

Cold waste fluid

Hot waste fluid

Hot fluid

Figure 9.1   Shell and tube heat exchanger

(a) *Shell and tube heat exchanger*: Shell and tube heat exchangers consist of a bundle of tubes inside a cylindrical shell through which two fluids flow, one through the tubes and the other through the shell (as shown in Figure 9.1). Heat is exchanged by conduction through the tube walls. Baffles are often used to direct fluid around the heat exchanger and also to provide structural support for the tubes.

(b) *Plate heat exchanger*: Plate heat exchangers consist of a large number of thin metal plates (usually stainless steel but sometimes titanium or nickel), which are clamped tightly together and sealed with gaskets (see Figure 9.2). The thin plates are profiled so that 'flow ways' are

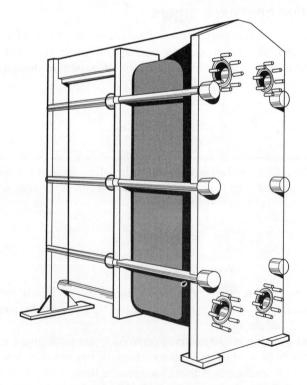

Figure 9.2   Plate heat exchanger

created between the plates when they are packed together, and a very large surface area is created across which heat transfer can take place. Ports located at the corners of the individual plate separate the 'hot' and 'cold' fluid flows and direct them to alternate passages so that no intermixing of the fluids occurs. The whole exchanger experiences a counter-flow pattern. The maximum operating temperature is usually about 130 °C if rubber sealing gaskets are fitted, but this can be extended to 200 °C if compressed asbestos fibre seals are used [1]. Plate heat exchangers have become popular in recent years because they are extremely compact, and can easily be expanded or contracted to accommodate future system modification.

(c) *Flat plate recuperator*: Flat plate recuperators consist of a series of metal (usually aluminium) plates separating 'hot' and 'cold' air or gas flows, sandwiched in a box-like structure (see Figure 9.3). The plates are sealed in order to prevent intermixing of the two fluid flows. They are often used in ducted air conditioning installations to reclaim heat from the exhaust air stream, without any cross contamination occurring.

9.3 Heat exchanger theory

The two most commonly used heat exchanger flow configurations are *counter flow* and *parallel flow*. These flow patterns are represented in Figures 9.4 and 9.5 respectively, along with their characteristic temperature profiles.

It should be noted that with the parallel flow configuration the 'hot' stream is always warmer than the 'cold' stream. With the counter flow configuration it is possible for the outlet temperature of the cold fluid to be higher than the outlet temperature of the hot fluid.

The general equations which govern the heat transfer in recuperative heat exchangers are as follows:

$$Q = \dot{m}_h\, c_h\, (t_{h1} - t_{h2}) = \dot{m}_c\, c_c\, (t_{c1} - t_{c2}) \tag{9.3}$$

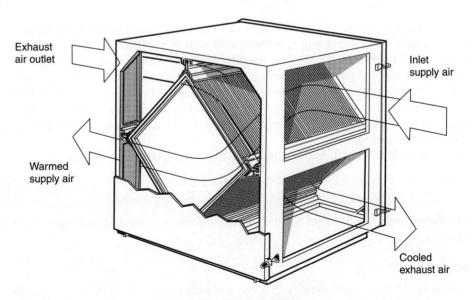

Figure 9.3   Flat plate recuperator

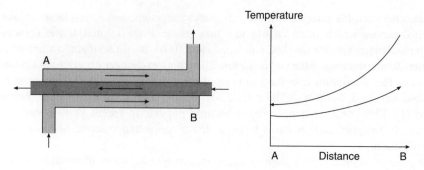

Figure 9.4   Counter flow heat exchanger

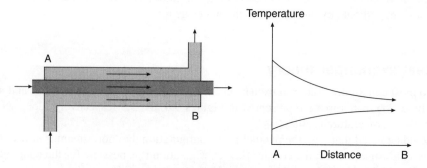

Figure 9.5   Parallel flow heat exchanger

and

$$Q = UA_o \, (\text{LMTD})K \tag{9.4}$$

where Q is the rate of heat transfer (W), $\dot{m}_h$ is the mass flow rate of hot fluid (kg/s), $\dot{m}_c$ is the mass flow rate of cold fluid (kg/s), c_h is the specific heat of hot fluid (J/kgK), c_c is the specific heat of cold fluid (J/kgK), t_{h1} and t_{h2} are the inlet and outlet temperatures of hot fluid (°C), t_{c1} and t_{c2} are the outlet and inlet temperatures of cold fluid (°C), U is the overall heat transfer coefficient (i.e. U value) (W/m²K), A_o is the outside surface area of heat exchanger (m²), LMTD is the logarithmic mean temperature difference (°C), and K is the constant which is dependent on the type of flow through the heat exchanger (e.g. $K = 1$ for counter-flow and parallel flow, and is therefore often ignored).

The logarithmic mean temperature difference (LMTD) can be determined by:

$$\text{LMTD} = \frac{\Delta t_1 - \Delta t_2}{\ln \, (\Delta t_1 / \Delta t_2)} \tag{9.5}$$

The following examples illustrate how the above equations can be used to design and analyse heat exchangers:

Example 9.1

A liquid waste stream has a flow rate of 3.5 kg/s and a temperature of 70 °C, with a specific heat capacity of 4190 J/kgK. Heat recovered from the hot waste stream is used to pre-heat boiler make-up water. The flow rate of the make-up water is 2 kg/s, its temperature is 10 °C and its specific heat capacity is 4190 J/kgK. The overall heat transfer coefficient of the heat exchanger is 800 W/m²K. If a make-up water exit temperature of 50 °C is required, and assuming that there are no heat losses from the exchanger, determine:

(i) The heat transfer rate;
(ii) The exit temperature of the effluent; and
(iii) The area of the heat exchanger required.

Solution

(i) Now:

$$Q = \dot{m}_c\, c_c\, (t_{c1} - t_{c2})$$
$$= 2 \times 4190 \times (50 - 10)$$
$$= 335\ 200\ W = 335.2\ kW$$

(ii) Now

$$\dot{m}_h\, c_h\, (t_{h1} - t_{h2}) = \dot{m}_c\, c_c\, (t_{c1} - t_{c2})$$
$$3.5 \times 4190 \times (70 - t_{h2}) = 2 \times 4190 \times (50 - 10)$$
$$t_{h2} = 47.14\ ^\circ C$$

(iii) Now, because the water outlet temperature is above the outlet temperature of the effluent, a counter-flow heat exchanger is required.

$$\text{LMTD} = \frac{\Delta t_1 - \Delta t_2}{\ln (\Delta t_1/\Delta t_2)}$$

$$= \frac{((70 - 50) - (47.14 - 10))}{\ln ((70 - 50)/(47.14 - 10))}$$

$$= 27.69\ ^\circ C$$

Now

$$Q = UA(\text{LMTD})$$

therefore

$$A = \frac{335\ 200}{800 \times 27.69} = 15.13\ \text{m}^2$$

Example 9.2

Consider the *counter flow* heat exchanger shown in Figure 9.6. Given the data below, determine the overall heat transfer rate for the heat exchanger.

Data:

  Length of heat exchanger = 2 m
  Internal radius of heat exchanger surface = 10 mm
  External radius of heat exchanger surface = 11 mm
  Thermal conductivity of heat exchanger surface = 386 W/mK
  Heat transfer coefficient of Fluid 1 = 50 W/m²K
  Heat transfer coefficient of Fluid 2 = 90 W/m²K

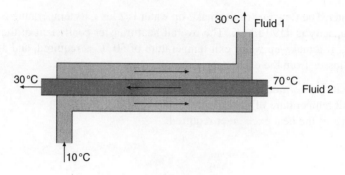

Figure 9.6   Heat exchanger

Solution

By combining eqns (8.26) and (8.28) it can be shown that the total thermal resistance, R_t, of the heat exchanger is:

$$R_t = \frac{1}{h \cdot A_1} \times \frac{\ln(r_2/r_1)}{2\pi k \cdot 1} \times \frac{1}{h \cdot A_2}$$

where k is the thermal conductivity of the pipe wall (W/mK), r_1 is the internal radius of the pipe (m), r_2 is the external radius of the pipe (m), 1 is the length of the pipe (m), h is the heat transfer coefficient (W/m²K), and A_1 and A_2 are the external and internal surface areas (m²)

and using eqn (8.30) the total heat transfer rate can be expressed as:

$$Q = \frac{\Delta t}{R_t} \text{(W)}$$

Now:

$$A_1 = 2\pi \times 0.011 \times 2 = 0.138 \text{ m}^2$$

and

$$A_2 = 2\pi \times 0.01 \times 2 = 0.126 \text{ m}^2$$

$$R_t = \frac{1}{90 \times 0.126} \times \frac{\ln(0.011/0.01)}{2\pi \times 386 \times 2} \times \frac{1}{50 \times 0.138}$$

therefore

$$R_t = 0.233 \text{ W/K}$$

and

$$\text{LMTD} = \frac{(70 - 30) - (30 - 10)}{\ln [(70 - 30)/(30 - 10)]} = 28.85 \text{ °C}$$

therefore

$$Q = \frac{28.85}{0.233} = 123.8 \text{ W}$$

9.3.1 Number of transfer units (NTU) concept

In some situations only the inlet temperatures and the flow rates of the *hot* and *cold* streams are known. Under these circumstances the use of the LMTD method results in a long and complex mathematical solution. To simplify such calculations the number of transfer units (NTU) method was developed [2, 3].

NTU is defined as the ratio of the temperature change of one of the fluids divided by the mean driving force between the fluids, and can be expressed as:

For the hot fluid:

$$NTU_h = \frac{(t_{h1} - t_{h2})}{(LMTD)K} = \frac{UA_o}{(\dot{m}c)_h} \tag{9.6}$$

For the cold fluid:

$$NTU_c = \frac{(t_{c1} - t_{c2})}{(LMTD)K} = \frac{UA}{(\dot{m}c)_c} \tag{9.7}$$

(NB: For counter flow and parallel flow heat exchangers the K term can be ignored)
Equations (9.6) and (9.7) are more commonly simplified to:

$$NTU = \frac{UA_o}{(\dot{m}c)_{min}} \tag{9.8}$$

where $(\dot{m}c)_{min}$ is the minimum thermal capacity (kW/K).

The ratio of the thermal capacities, R, is defined as:

$$R = \frac{(\dot{m}c)_{min}}{(\dot{m}c)_{max}} \tag{9.9}$$

If both fluids in the heat exchanger have the same thermal capacity then $R = 1$. At the other extreme when one of the fluids has an infinite thermal capacity, as in the case of an evaporating vapour, then $R = 0$.

Another useful concept is the *effectiveness*, E, of a heat exchanger. Effectiveness can be defined as the actual heat transfer divided by the maximum possible heat transfer across the heat exchanger, and can be expressed as:

$$E = \frac{Q}{Q_{max}} = \frac{Q}{(\dot{m}c)_{min}(t_{hmax} - t_{cmin})} \tag{9.10}$$

It is possible to derive the relationship between E, NTU and R for a variety of heat exchanger applications. The mathematical expressions for some of the more common applications are given below.

(i) Parallel flow:

$$E = \frac{1 - e^{[-NTU\,(1+R)]}}{1 + R} \tag{9.11}$$

(ii) Counter flow:

$$E = \frac{1 - e^{[-NTU\,(1-R)]}}{1 - Re^{[-NTU\,(1-R)]}} \tag{9.12}$$

If R = 1 then this expression simplifies to:

$$E = \frac{NTU}{1 + NTU} \tag{9.13}$$

(iii) Heat exchanger with condensing vapour of boiling liquid on one side:

$$E = 1 - e^{[-NTU]} \qquad (9.14)$$

The NTU method for heat exchanger analysis is illustrated in Example 9.3.

Example 9.3

A contaminated water stream from a factory building has a temperature of 80 °C, and has a flow rate of 6 kg/s and a specific heat capacity of 4.19 kJ/kg K. The incoming water supply to the manufacturing process is at 10 °C, and has a flow rate of 7 kg/s and a specific heat capacity of 4.19 kJ/kg K. It is proposed to install a counter flow heat exchanger to recover the waste heat. If the heat exchanger has an overall area of 30 m² and an overall heat transfer coefficient of 800 W/m²K (assuming that there are no heat losses from the heat exchanger), determine:

(i) The effectiveness of the heat exchanger;
(ii) The heat transfer rate; and
(iii) The exit temperature of the incoming water stream leaving the heat exchanger.

Solution

Now

$$(\dot{m}c)_{min} = 6 \times 4.19 = 25.14 \text{ kW/K}$$

and

$$(\dot{m}c)_{max} = 7 \times 4.19 = 29.33 \text{ kW/K}.$$

therefore

$$R = \frac{25.14}{29.33} = 0.857$$

and

$$NTU = \frac{30 \times 800}{25.14 \times 1000} = 0.955.$$

(i) Therefore

$$E = \frac{1 - e^{[-0.955\,(1-0.857)\,]}}{1 - 0.857\,e^{[-0.955\,(1-0.857)\,]}}$$

$$E = 0.506$$

(ii) Now

$$E = \frac{Q}{(\dot{m}c)_{min}(t_{hmax} - t_{cmin})}$$

therefore:

$$Q = 0.506 \times 25.14 \times [80 - 10] = 890.46 \text{ kW}$$

(iii) Therefore:

$$890.46 = 29.33 \times [t_{off} - 10]$$

therefore:

$$t_{off} = 40.4 \text{ °C}$$

9.4 Run-around coils

When two recuperative heat exchangers are linked together by a third fluid which transports heat between them, the system is known as a run-around coil. Run-around coils are often employed to recover waste heat from exhaust air streams and to pre-heat incoming supply air, thus avoiding the risk of cross-contamination between the two air streams. Such a system is shown in Figure 9.7. Run-around coils usually employ a glycol/water mixture as the working fluid which avoids the risk of freezing during the winter.

Run-around coils have the advantage that they can be used in applications where the two fluid streams are physically too far apart to use a recuperative heat exchanger. Whilst this feature is usually considered advantageous it does result in increased energy consumption since a pump is introduced into the system, and may also result in heat loss from the secondary fluid. This makes it important to insulate the pipework circuit, otherwise the overall effectiveness of the system will become unacceptably low. Despite these drawbacks, when compared with many other methods of recovering waste heat, run-around coils are relatively inexpensive to install since they utilize standard air/water heating coils.

In the case of the system shown in Figure 9.7 the thermal capacity ($\dot{m}c$) of the *cold* fluid is equal to that of the *hot* fluid since the two heat exchangers are identical. Therefore:

$$(\dot{m}c)_h = (\dot{m}c)_c = (\dot{m}c)_s \tag{9.15}$$

Where $(\dot{m}c)_s$ is the thermal capacity of the secondary fluid (kW/K).

Consequently, it can be shown that:

$$t_{s1} = \frac{(t_{h1} + t_{c1})}{2}$$

and

$$t_{s2} = \frac{(t_{h2} + t_{c2})}{2}$$

Where t_{s1}, t_{s2} are the flow and return temperatures of the secondary fluid (°C), t_{h1}, t_{h2} are the temperatures of the hot fluid stream before and after heat exchanger (°C), and t_{c2}, t_{c1} are the temperatures of the cold fluid stream before and after heat exchanger (°C).

Also, the overall heat transfer can be defined by

$$Q = (UA)_o (t_{h1} - t_{c1})$$

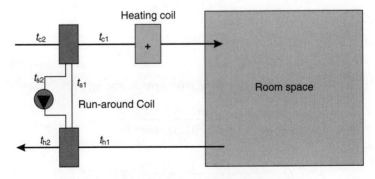

Figure 9.7   Run-around coil system

and

$$Q = (UA)_h (t_{h1} - t_{s1})$$

$$= (UA)_h \left(t_{h1} - \frac{(t_{h1} + t_{c1})}{2} \right)$$

where $(UA)_o$ is the product of U and A for the whole run-around coil (W/K), and $(UA)_h$ is the product of U and A for the heat exchanger in the hot stream (W/K).
Therefore:

$$(UA)_o (t_{h1} - t_{c1}) = (UA)_h \frac{(t_{h1} - t_{c1})}{2}$$

so

$$Q = \frac{(UA)_h (t_{h1} - t_{c1})}{2}$$

and since

$$Q = (\dot{m}c)_c \times (t_{c1} - t_{c2})$$

it can be shown that:

$$Q = \frac{(UA)_h (t_{h1} - t_{c2})}{2 + [(UA)_h/(\dot{m}c)_c]} \tag{9.16}$$

Example 9.4

A run-around coil is applied to a heating and ventilation system as shown in Figure 9.7. Air is supplied to the room space at 28 °C and leaves at 20 °C. The outside air temperature is −1 °C. The supply air to the space has a mass flow rate of 3 kg/s and a mean specific heat capacity of 1.012 kJ/kgK. The specific heat of the secondary fluid is 3.6 kJ/kgK, and:

$$(UA)_c = (UA)_h = 5 \text{ kW/K}$$

Given this information, determine:

(i)   The required mass flow rate of the secondary fluid;
(ii) The temperature of the air entering the supply air heating coil; and
(iii) The percentage energy saving achieved by using the run-around coil.

Solution

(i) From eqn (9.15) it can be seen that:

$$(\dot{m}c)_c = (\dot{m}c)_s$$

therefore:

$$\dot{m}_s = \frac{3 \times 1.012}{3.6} = 0.843 \text{ kg/s}$$

(ii) Now:

$$Q = \frac{(UA)_h (t_{h1} - t_{c2})}{2 + [(UA)_h/(\dot{m}c)_c]}$$

therefore:

$$Q = \frac{5 \times [20 - (-1)]}{2 + [5/(3 \times 1.012)]} = 28.79 \text{ kW}$$

and since

$$t_{c1} = t_{c2} + \frac{Q}{(\dot{m}c)_c}$$

therefore

$$t_{c1} = -1 + \frac{28.79}{(3 \times 1.012)} = 8.48 \, °C.$$

(iii) With the run-around coil

$$Q = 3 \times 1.012 \times [28 - 8.48] = 59.263 \, kW$$

Without run-around coil

$$Q = 3 \times 1.012 \times [28 - (-1)] = 88.044 \, kW$$

therefore

$$\text{Percentage saving} = \frac{(88.044 - 59.263)}{88.044} \times 100 = 32.7\%$$

While it is relatively simple to derive an expression for the heat transfer of a run-around coil when the thermal capacities of the fluids are equal, it becomes much more complex when the thermal capacities of the two fluids are different, and the heat exchangers are also different. However, this problem can be overcome by using the NTU method.

It can be shown that for a run-around coil

$$\frac{1}{(UA)_o} = \frac{1}{(UA)_h} + \frac{1}{(UA)_c} \tag{9.17}$$

and from eqn (9.8)

$$\text{NTU} = \frac{UA_o}{(\dot{m}c)_{min}},$$

therefore

$$\text{NTU} = \frac{UA_h \times UA_c}{(\dot{m}c)_{min}(UA_h + UA_c)} \tag{9.18}$$

Example 9.5 illustrates how the NTU method can be applied to a run-around coil problem.

Example 9.5

It is intended that a run-around coil be installed to recover waste heat from a flue gas stream at 250 °C, and to pre-heat a water stream at 10 °C. The flue gas has a mass flow rate of 4 kg/s and that of the water is 2 kg/s. The individual heat exchangers used in the system are of a counter flow type. Given the following data determine:

(i) The overall effectiveness of the run-around coil; and
(ii) The exit temperature of the water stream.

Data:

Specific heat capacity of flue gas = 1.2 kJ/kgK
Specific heat capacity of water = 4.19 kJ/kgK
UA for the flue gas heat exchanger = 5 kW/K
UA for the water heat exchanger = 18 kW/K

Solution

(i) Now

$$(\dot{m}c)_{min} = 4 \times 1.2 = 4.8 \, kW/K$$

and

$$(\dot{m}c)_{max} = 2 \times 4.19 = 8.38 \text{ kW/K}$$

therefore

$$R = \frac{4.8}{8.38} = 0.573$$

and

$$NTU = \frac{5 \times 18}{4.8 \times (5 + 18)} = 0.815$$

and from eqn (9.13)

$$E = \frac{1 - e^{[-0.815\,(1-0.573)\,]}}{1 - 0.573e^{[-0.815\,(1-0.573)\,]}}$$

$$= 0.494$$

(ii) Now

$$E = \frac{Q}{(\dot{m}c)_{min}(t_{hmax} - t_{cmin})}$$

therefore

$$Q = 0.494 \times 4.8 \times [250 - 10] = 569.1 \text{ kW}$$

therefore

$$569.1 = 8.38 \times [t_{off} - 10]$$

therefore

$$t_{off} = 77.9 \text{ °C}$$

9.5 Regenerative heat exchangers

In a regenerative heat exchanger a matrix of material is alternately passed from a hot fluid to a cold fluid, so that heat is transferred between the two in a cyclical process. The most commonly used type of regenerative heat exchanger is the thermal wheel, which has a matrix of material mounted on a wheel, which slowly rotates at approximately 10 revolutions per minute, through hot and cold fluid streams (as shown in Figure 9.8). The major advantage of a thermal wheel is that there is a large surface area to volume ratio resulting in a relatively low cost per unit surface area.

The matrix material in a thermal wheel is usually an open structured metal, such as knitted stainless steel or aluminium wire, or corrugated sheet aluminium or steel [1]. For use at higher temperatures honeycomb ceramic materials are used. Although thermal wheels are usually employed solely to recover sensible heat, it is possible to reclaim the enthalpy of vaporization of the moisture in the 'hot' stream passing through a thermal wheel. This is achieved by coating a non-metallic matrix with a hygroscopic or desiccant material such as lithium chloride [1].

Thermal wheels do have the major disadvantage that there is the possibility of cross-contamination between the air streams. This can be reduced considerably by ensuring that the cleaner of the two fluids is maintained at the highest pressure, and by using a purging device. Most thermal wheels incorporate a purge unit which allows a small proportion of the supply air to

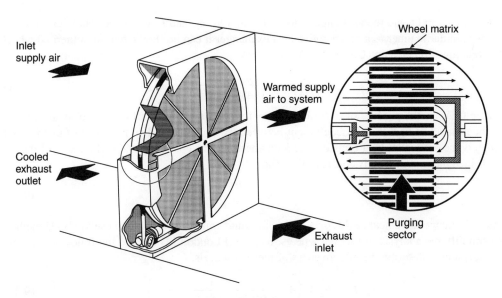

Figure 9.8   Thermal wheel

flush the contaminants from the wheel, thus keeping cross-contamination to a minimum (e.g. less than 0.04%) [1].

Thermal wheels are often used to recover heat from room ventilation systems such as that shown in Figure 9.9. In this type of application the thermal efficiency, η_t can be defined by:

$$\eta_t = \frac{t_2 + t_1}{t_3 + t_1} \tag{9.19}$$

similarly the overall (total energy) efficiency, η_x can be expressed:

$$\eta_x = \frac{h_2 + h_1}{h_3 + h_1} \tag{9.20}$$

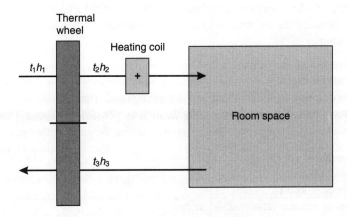

Figure 9.9   Thermal wheel application

where t_1, t_2, and t_3 are the air temperatures (°C), and h_1, h_2, and h_3 are the air enthalpies (°C).

In a similar manner to a recuperative heat exchanger it can be shown that for a thermal wheel the relationship between (UA) and (hA) is:

$$\frac{1}{(UA)_o} = \frac{1}{(hA)_h} + \frac{1}{(hA)_c} \qquad (9.21)$$

where $(UA)_o$ is the product of overall heat transfer coefficient and surface area of matrix, $(hA)_h$ is the product of heat transfer coefficient between the hot fluid and surface area of matrix, and $(hA)_c$ is the product of heat transfer coefficient between the cold fluid and surface area of matrix and since the matrix area is constant, therefore:

$$U = \frac{h}{2} \qquad (9.22)$$

As with recuperative heat exchangers and run-around coils it is possible to use the NTU method to simplify the analysis of thermal wheels. Kays and London [4] derived the following empirical formula to describe the effectiveness of thermal wheels.

$$E = E_c \times \left(1 - \frac{1}{9[(\dot{m}c)_M/(\dot{m}c)_{min}]^{1.93}}\right) \qquad (9.23)$$

where

$$(\dot{m}c)_M = N \cdot M \cdot c_M \qquad (9.24)$$

N is the wheel revolutions per second, M is the mass of the matrix (kg), c_M is the specific heat capacity of matrix material (kJ/kgK), and;

$$E_c = \frac{1 - e^{[-NTU(1-R)]}}{1 - Re^{[-NTU(1-R)]}} \quad \text{where} \quad R = (\dot{m}c)_{min}/(\dot{m}c)_{max}$$

or

$$E_c = \frac{NTU}{1 + NTU} \quad \text{when} \quad R = 1$$

Example 9.6

The exhaust air from a factory building is at a temperature of 35 °C and has a flow rate of 6 kg/s and a specific heat capacity of 1.025 kJ/kg K. The incoming fresh air to the building is at −1 °C, and has a flow rate of 7 kg/s and a specific heat capacity of 1.025 kJ/kg K. It is proposed to insert a thermal wheel between the air streams to recover the sensible waste heat. Given the following information determine:

(i)   The effectiveness of the thermal wheel
(ii) The heat transfer rate
(iii) The exit temperature of the fresh air leaving the thermal wheel
(iv) The exit temperature of the fresh air leaving the thermal wheel if its rotational speed is doubled.

Data:

    Wheel diameter = 1.2 m
    Wheel depth = 0.4 m
    Mass of wheel = 140 kg
    Surface area to volume ratio = 2500 m^2/m^3
    Specific heat of matrix material = 1.3 kJ/kgK

Wheel speed = 8 rev/min
Heat transfer coefficient for each = 35 W/m²K
air stream

Solution

$$\text{Face area of wheel} = \frac{\pi \times 1.2^2}{4} = 1.12 \text{ m}^2$$

$$\text{Volume of wheel} = 1.13 \times 0.4 = 0.452 \text{ m}^3$$

(i)
$$A = 0.452 - 2500 = 1130 \text{ m}^2$$

$$(\dot{m}c)_{\text{min}} = 6 \times 1.025 = 6.15 \text{ kW/K}$$

$$(\dot{m}c)_{\text{max}} = 7 \times 1.025 = 7.175 \text{ kW/K}$$

therefore

$$R = \frac{6.15}{7.175} = 0.857$$

and from eqn (9.21);

$$U = \frac{h}{2} = \frac{35}{2} = 17.5 \text{ W/m}^2\text{K}$$

therefore

$$\text{NTU} = \frac{UA_o}{(\dot{m}c)_{\text{min}}} = \frac{1130 \times 17.5}{6.15 \times 1000} = 3.215$$

therefore

$$E_c = \frac{1 - e^{[-3.215\,(1-0.857)]}}{1 - 0.857\, e^{[-3.215\,(1-0.857)]}} = 0.803$$

and

$$(\dot{m}c)_M = N \cdot M \cdot c_M = \frac{8}{60} \times 140 \times 1.3 = 24.27 \text{ kW/K}$$

therefore

$$E = 0.803 \times \left(1 - \frac{1}{9[24.27\,/\,6.15]^{1.93}}\right)$$

$$= 0.797$$

(ii) Now

$$E = \frac{Q}{(\dot{m}c)_{\text{min}}\,(t_{\text{hmax}} - t_{\text{cmin}})}$$

therefore

$$Q = 0.797 \times 6.15 \times [35 - (-1)] = 176.46 \text{ kW}$$

(iii) therefore

$$176.46 = 7.175 \times [t_{off} - (-1)]$$

therefore

$$t_{off} = 23.6\,°C$$

(iv) If $N = 2 \times 8 = 16$ rev/min, then

$$(\dot{m}c)_M = N \cdot M \cdot c_M = \frac{16}{60} \times 140 \times 1.3 = 48.53\ kW/K$$

therefore

$$E = 0.803 \times \left(1 - \frac{1}{9[48.53/6.15]^{1.93}}\right)$$
$$= 0.801$$

therefore

$$Q = 0.801 \times 6.15 \times [35 - (-1)] = 177.43\ kW$$

therefore

$$177.34 = 7.175 \times [t_{off} - (-1)]$$

therefore

$$t_{off} = 23.7\,°C$$

From this it can be seen that there is very little benefit to be gained from doubling the rotational speed of the thermal wheel.

9.6 Heat pumps

A heat pump is essentially a vapour compression refrigeration machine which takes heat from a low temperature source such as air or water, and upgrades it to be used at a higher temperature. Unlike a conventional refrigeration machine, the heat produced at the condenser is utilized and not wasted to the atmosphere. Figure 9.10 shows a simple vapour compression heat pump, together with the relevant pressure/enthalpy diagram.

The performance of the vapour compression refrigeration cycle is quantified by the *coefficient of performance* (COP), which can be expressed as:

for a refrigeration machine:

$$COP_{ref} = \frac{useful\ refrigeration\ output}{net\ work\ input}$$

for a heat pump:

$$COP_{hp} = \frac{useful\ heat\ rejected\ from\ cycle}{net\ work\ input}$$

The COP of the vapour compression cycle is usually expressed in terms of a ratio of enthalpy differences, hence the COP of a refrigeration machine can be expressed as follows (referring to Figure 9.10):

$$COP_{ref} = \frac{h_1 - h_4}{h_2 - h_1}$$
(9.25)

where:

$$h = \text{specific enthalpy of refrigerant (kJ/kg)}$$

So, for a heat pump:

$$COP_{hp} = \frac{h_2 - h_3}{h_2 - h_1}$$
(9.26)

from this it can be shown that:

$$COP_{hp} = COP_{ref} + 1$$
(9.27)

For an ideal heat pump the maximum possible COP is given by the Carnot cycle expression:

$$COP_{hp} = \frac{T_c}{T_c - T_e}$$
(9.28)

where T_e is the evaporating absolute temperature (K), and T_c is the condensing absolute temperature (K).

In practice the Carnot COP shown above can never be achieved, but the Carnot equation shows that the greater the difference between T_c and T_e the lower the COP of the heat pump. Heat pumps are therefore well suited to applications where the evaporating and condensing temperatures are close together, which is the case when recovering heat from exhaust air in heating and air conditioning applications. As a result, heat pumps are often used in air conditioning applications. They are also popular in applications where there is a need for both dehumidification and heating, such as in warehouses where the occurrence of a high humidity may cause condensation problems and result in the destruction of valuable stock.

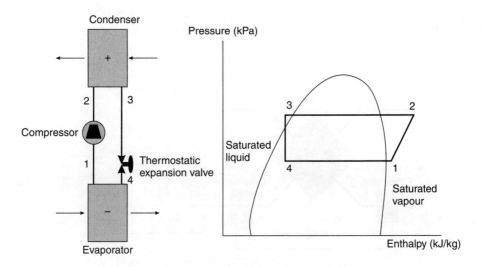

Figure 9.10   Vapour compression heat pump

Swimming pool buildings are particularly well suited to the application of heat pumps. In swimming pools the air leaving the pool hall is very humid and contains large amounts of latent heat bound up in the water vapour. Heat pumps are particularly well suited to recovering the enthalpy of vaporization from the moisture in the exhaust air. A typical example of the heat pump used in combination with a flat plate heat recuperator is shown in Figure 9.11. In this application sensible heat is taken from the swimming pool exhaust air by the flat plate recuperator and used to pre-heat the supply air stream. The evaporator of the heat pump then dehumidifies the exhaust air stream and recovers the latent heat bound up in the water vapour. The heat pump then rejects this heat (plus the 'work' input by the compressor) through the condenser, and thus heats the supply air to the pool.

Example 9.7

The heat pump shown in Figure 9.11, operates on refrigerant HCFC 22. Given the following data, calculate:

(i) The COP of the heat pump
(ii) The electrical energy consumed for each kW of heat produced

Data:

Condensing temperature = 50 °C
Evaporating temperature = 10 °C
Vapour temperature (leaving compressor) = 80 °C
Liquid temperature (leaving condenser) = 40 °C
Combined electrical and mechanical efficiency of motor = 90%

Solution

Using a pressure enthalpy chart (see Appendix 2), or by using thermodynamic tables for HCFC 22 it is possible to plot the refrigeration process as follows:

(i)
$$COP_{hp} = \frac{h_2 - h_3}{h_2 - h_1}$$

$$= \frac{346 - 150}{346 - 315} = 6.323$$

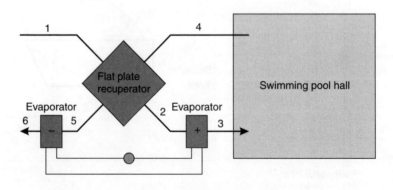

Figure 9.11   Heat recovery system for a swimming pool building

$$\text{(ii)} \qquad \text{Electricity consumption per kW of produced} = \frac{1}{6.323 \times 0.9} = 0.176 \text{ kW}$$

Example 9.8

For the heat pump installation shown in Figure 9.11, calculate:

(i) The heat output of the heat pump
(ii) The mass flow rate of refrigerant required in the heat pump circuit
(iii) The power input required to the electric motor
(iv) The specific enthalpy of the air leaving the evaporator coil

Data:

The mass flow rate of supply air = 6 kg/s and
Condition of air leaving pool hall = 29 °C and 70% saturation
Temperature of air supplied to pool hall = 34 °C
Outside air condition = −3 °C and 100% saturation
Effectiveness of flat plate recuperator = 0.7

Solution

Consider first the fresh outside air entering the system and passing through the flat plate recuperator. It enters the system at −3 °C and 100% saturation; from a psychrometric chart (see Appendix 3) or from psychrometric tables, it can be determined that the moisture content of incoming air stream is 0.0029 kg/kg (dry air) and its specific enthalpy is 4.2 kJ/kg.

Now

$$\text{The effectiveness of a flat plate recuperator} = \frac{\text{Heat transferred}}{\text{Max. theoretical heat transferred}}$$

Therefore, for the supply air stream, if the maximum theoretical heat transfer occurred, then it would be heated from −1 °C to 29 °C at a constant moisture content of 0.0029 kg/kg. From a psychrometric chart or tables, the specific enthalpy of air at 29 °C and 0.0029 kg/kg is 36.6 kJ/kg.

Therefore

The maximum theoretical heat transfer = 36.6 − 4.2

therefore

$$\text{Effectiveness} = \frac{\text{Heat transferred}}{(37.2 - 4.2)}$$

therefore

$$\text{Heat transferred } (h_2 - h_1) = 0.7 \, (36.6 - 4.2) = 22.68 \text{ kJ/kg}$$

therefore

$$h_2 - 4.2 = 22.68 \text{ kJ/kg}$$
$$h_2 = 22.68 + 4.2 = 26.88 \text{ kJ/kg}$$

At a moisture content of 0.0029 kg/kg, h_2 equates to an air temperature of 19.3 °C. The heat pump condenser therefore has to raise the supply air temperature from 19.3 °C to 34 °C, at which temperature the specific enthalpy is 41.6 kJ/kg.

(i) Therefore

$$Q_{cond} = \dot{m}_{air} \times (h_3 - h_2)$$
$$= 6 \times (41.6 - 26.88) = 88.32 \text{ kW}$$

and from Example 9.7 it can be seen that for the condenser

(ii) $$Q_{cond} = \dot{m}_{ref} \times (346 - 150)$$

therefore

$$\dot{m}_{ref} = \frac{88.32}{346 - 150} = 0.451 \text{ kg /s}$$

(iii) Therefore

$$\text{Electric power input to compressor} = \frac{0.451 \, (346 - 315)}{0.9} = 15.534 \text{ kW}$$

(iv) Considering now the exhaust air stream through the flat plate recuperator

$$Q_{fpr} = \dot{m}_{air} \times (h_2 - h_1) = \dot{m}_{air} \times (h_4 - h_5)$$

therefore

$$Q_{fpr} = 6 \times 22.68 = 136.08 \text{ kW}$$

Now, the air leaving the pool hall has a moisture content of 0.018 kg/kg and specific enthalpy (h_4) of 75.1 kJ/kg. therefore:

$$h_5 = h_4 - \frac{Q_{fpr}}{\dot{m}_{air}}$$

therefore

$$h_5 = 75.1 - \frac{136.08}{6} = 52.42 \text{ kJ/ kg}$$

and from Example 9.7;

$$Q_{evap} = 0.451 \times (315 - 150) = 74.415 \text{ kW}$$

and

$$Q_{evap} = \dot{m}_{air} \times (h_5 - h_6)$$

therefore

$$h_6 = 52.42 - \frac{74.414}{6} = 40.02 \text{ kJ/kg}$$

Many manufacturers produce machines which have the dual ability to act as both a refrigeration machine and a heat pump. These machines have twin condensers; an air cooled one for normal operation, and a water cooled one for the heat pump mode. They are often installed in buildings and act as air conditioning chillers. When operating in the heat pump mode the waste heat from the condenser is recovered and used to produce the domestic hot water for the building. This at first sight would appear to be a classic energy conservation measure. However, such 'energy saving' measures should be treated with caution since in order to produce the domestic hot water it may be necessary to raise the condensing pressure considerably, with the result that the COP may be significantly reduced. When it is also considered that the unit price of electricity is usually 3 to 4 times that of gas, then the adoption of such a dual purpose machine may not be quite as advantageous as it appeared originally.

References

1. Cornforth, J. R. (1992). *Combustion Engineering and Gas Utilisation* (Chapter 7). E & FN Spon.
2. Eastop, T. D. and Croft, D. R. (1990). *Energy Efficiency for Engineers and Technologists* (Chapter 5). Longman.
3. Incropera, F. P. and De Witt, D. P. (1990). *Fundamentals of heat and mass transfer* (Chapter 11). John Wiley & Sons.
4. Kays, W. M. and London, A. L. (1984). *Compact Heat Exchangers*. Mc Graw-Hill.

Bibliography

Brookes, G. (1985). Assessing the scope for heat recovery. *Energy Manager's Workbook*, **2**, (Chapter 6). Energy Publications.

Cornforth, J. R. (1992). *Combustion Engineering and Gas Utilisation* (Chapter 7). E & FN Spon.

Eastop, T. D. and Croft, D. R. (1990). *Energy Efficiency for Engineers and Technologists* (Chapter 5). Longman.

McQuiston, F. C. and Parker, J. D. (1994). *Heating, ventilating, and air conditioning* (Chapter 14). John Wiley & Sons.

Ozisik, M. N. (1985). *Heat transfer: a basic approach* (Chapter 11). Mc Graw-Hill.

Stoecker, W. F. and Jones, J. W. (1982). *Refrigeration and air conditioning* (Chapter 2). Mc Graw-Hill, 2nd edition.

Thumann, A. and Mehta, D. P. (1997). *Handbook of energy engineering* (Chapter 5). The Fairmont Press (Prentice Hall), 4th edition.

10

Combined heat and power

This chapter investigates the subject of combined heat and power (CHP). The general nature of combined heat and power systems is discussed and the economic benefits appraised. In particular, CHP plant sizing strategies are evaluated and example design calculations presented.

10.1 The combined heat and power concept

From an energy point of view, the generation of electricity in thermal power stations is an extremely wasteful process. Most conventional thermal power stations exhibit efficiencies in the range 30–37% [1], while the newer combined cycle gas turbine stations still only achieve efficiencies in the region of 47% [1]. This means that over 50% of the primary energy consumed in the generation process is wasted and not converted into delivered electricity. This wasted energy is converted to heat which is ultimately rejected to the environment. The generation process also liberates considerable amounts of CO_2 into the atmosphere. It has been calculated that in the UK 0.43 kg of CO_2 is liberated for every 1 kWh of electrical energy delivered (2001 data) [2].

One easy way to appreciate the inefficiency of the electricity generation cycle is to consider the theoretical maximum efficiency of the process. The Carnot principle shows that the theoretical maximum thermal efficiency of any heat engine cycle can be determined by:

$$\eta_{carnot} = 1 - \frac{T_2}{T_1}$$

where T_1 is the the maximum temperature available (K), T_2 is the the lowest temperature available (K).

For example, if the maximum temperature in a cycle is 1450 K and the cooling water minimum temperature is 285 K, then the maximum possible efficiency of the cycle is:

$$\eta_{carnot} = 1 - \frac{285}{1450} = 0.803 \text{ (or } 80.3\%)$$

The fact that this level of efficiency is not achieved in practice is due to the high degree of irreversibility in the process. Consequently, the efficiencies achieved in power stations are very much lower than the theoretical Carnot efficiency and are dependent on the type of prime mover used.

The low operating efficiencies achieved during the electricity generation process result in a great amount of energy being lost in the form of waste heat. Given the Earth's dwindling energy resources this is not a very satisfactory arrangement. It would be much better to collect the waste heat from the generation process and use it to heat buildings. By combining the *electrical generation* and *heat production* processes it is possible to produce a highly efficient system which makes good use of primary energy. It is the combination of the *electrical generation* and *heat production* processes which is the basis of the CHP, or *cogeneration*, concept. In a typical CHP installation heat exchangers are used to reclaim waste heat from exhaust gases and other sources during the electricity generation process. In this way it is possible to achieve overall efficiencies in the region of 80%, if a system is correctly optimized [3].

CHP systems vary in size from large 'power stations' serving whole cities to small micro-CHP units serving individual buildings. The larger CHP systems tend to use gas or steam turbines, while smaller systems generally use internal combustion engines converted to run on natural gas. During the electricity generation process waste heat is recovered from the exhaust gases, or used steam, and, in the case of micro-CHP systems, also from the engine jacket. Large cogeneration systems often use recovered heat to produce hot water for use in district heating schemes, while micro-CHP systems are generally used to heat single buildings.

CHP schemes enable electricity to be generated locally, and eliminate much of the wastage of heat which normally occurs in conventional power plants. Through the use of CHP it is possible to:

- Improve national energy efficiency and preserve non-renewable energy reserves. This is particularly important for nations which have limited fossil fuel resources and which are dependent on imported energy.
- Reduce the cost of transporting electrical energy. The transportation of electricity over long distances involves the construction of expensive transmission networks (consisting of cables, pylons, transformers and switchgear). The need for these is reduced by the use of locally-based CHP schemes. Localized CHP schemes also save energy because they avert the need to transport electricity over long distances. There is a 4–8% energy loss during the transportation of electricity over long distances.
- Reduce the amount atmospheric pollution produced, due to more efficient fuel conversion.

Although CHP has many potential benefits, there are a number of problems associated with it, which have inhibited its widespread use.

- CHP plant requires considerable capital expenditure. This necessitates a full financial appraisal of future energy demands, fuel prices and maintenance costs. Such an appraisal may only be accurate in the short term, with the result that organizations often 'play safe' and rely on con-ventional systems with which they are familiar.
- There must be a demand for the heat from any proposed CHP plant. Although in most applications it is possible to fully utilize the electricity produced by CHP plant, it is often much more difficult to utilize the heat which is produced. Most building types do not have the all-year-round demand for heat which is required to successfully employ a CHP plant. On the contrary many building types require cooling for large parts of the year.
- Back-up plant is often required in CHP installations, in order to ensure security of supply of electricity and heat. This 'back-up' plant adds to the capital cost of the installation.

Given the considerable capital expenditure associated with CHP schemes it is essential that any proposed CHP application be carefully evaluated to determine its suitability. It should be remembered with caution that there are many so-called *energy saving* schemes which have proved to be expensive liabilities.

10.2 CHP system efficiency

It is possible to illustrate the energy saving merits of CHP systems by comparing the primary energy consumption of a typical micro-CHP plant with that consumed by a conventional system in which heat is produced in a boiler and electrical power is purchased from a utility company. Example 10.1 presents the energy balance for the two alternative systems.

Example 10.1

A building has an electrical power requirement of 80 kWe (i.e. 80 kW of electrical power) and a heat load of 122 kW. The owners of the building are considering installing a micro-CHP unit which utilizes an internal combustion engine converted to run on natural gas. Compare the primary energy consumption and unit energy costs of the CHP scheme compared with a conventional separate system.

Data:

    Efficiency of the conventional electricity supply process = 35%
    Efficiency of conventional boiler plant = 70%
    Mechanical efficiency of CHP unit = 32%
    Efficiency of CHP electricity generator = 95%
    Heat recovery efficiency of CHP unit = 68.16%
    Unit cost of gas = 0.9 p/kWh
    Unit cost of electricity = 5.0 p/kWh

Solution

The two options are considered as follows:
Option 1: conventional system:

$$\text{Primary fuel power input to generate electicity} = \frac{80}{0.35} = 228.6 \text{ kW}$$

and

$$\text{Power input to boilers} = \frac{122}{0.70} = 174.3 \text{ kW.}$$

Therefore

$$\text{Total primary power input} = 228.6 + 174.3 = 402.9 \text{ kW}$$

therefore

$$\text{Overall system efficiency} = \frac{80 + 122}{402.9} \times 100 = 50.1\%$$

and

$$\text{Energy cost for 1 hour's operation} = \frac{(80 \times 5.0) + (174.3 \times 0.9)}{100} = £5.57$$

Option 2: CHP System:

$$\text{Fuel power input to CHP unit} = \frac{80}{0.32 \times 0.95} = 263.2 \text{ kW}$$

The waste heat produced by the CHP unit is passed through a heat exchanger with an efficiency of 68.16%, therefore:

$$\text{Recoverable heat power} = (263.2 \times (1 - 0.32)) \times 0.6816 = 122.0 \text{ kW}$$

Therefore

$$\text{Overall system efficiency} = \frac{80 + 122}{263.2} \times 100 = 76.7\%$$

and

$$\text{Energy cost for 1 hour's operation} = (263.2 \times 0.9) = £2.37$$

Example 10.1 clearly shows that there are large potential energy cost savings to be gained through utilizing CHP in buildings.

10.3 CHP systems

CHP systems can range from small 'micro' installations, designed to serve the needs of a single building, to large systems which satisfy the heating and electrical power requirements of whole towns. Micro-CHP systems utilizing internal combustion engines tend to be used in applications where electrical demand does not exceed 1 MWe. Gas turbines are popular on larger installations, while steam turbines are often used on the largest schemes.

10.3.1 Internal combustion engines

Internal combustion engines are often used to drive small micro-CHP systems. Mechanical power from this type of engine is used to drive a generator and heat is recovered from the engine exhaust, jacket water and lubricating oil. Micro-CHP units typically operate in the range 15 kWe to 1 MWe electrical output. Modified automotive derived engines are the most widely used systems up to 200 kWe electrical output, whereas more rugged stationary industrial engines are generally used for higher outputs [3]. The automotive engines used are generally modified lorry engines, which are converted to run on gas. These engines usually operate at a much slower and constant speed, typically 1500 rpm, than normal automotive engines. The engine life of a typical CHP prime mover is thus considerably longer than that of a typical automotive engine. Spark ignition gas engines tend to exhibit a heat to power ratio around 1.7:1 [3], whereas compression ignition diesel engines have heat to power ratios nearer 1:1.

10.3.2 Gas turbines

Where a natural gas supply is available, gas turbines are often used as the prime mover for larger CHP systems. Gas turbines have a relatively low capital cost and are reliable. The peak load mechanical efficiency of gas turbines is around 30%, which gives an optimum heat to power ratio of around 3:1 [4]. However, under part-load conditions efficiency can be substantially reduced. Gas turbines are usually fuelled by natural gas, but oil and pulverized coal have also been successfully employed.

A typical gas turbine CHP arrangement is shown in Figure 10.1. An air compressor, turbine and generator are mounted on a single shaft, with the turbine being the prime mover. Gas turbines employ an open cycle in which air is drawn into a compressor and compressed to a high pressure before being introduced into a combustion chamber where natural gas is burnt. On leaving the combustion chamber the pressurized combustion gases are forced at temperatures between 900 °C to 1200 °C [4] through a turbine, which in turn rotates a generator. On exiting the turbine, the hot combustion gases, at 450 °C to 550 °C [4], pass through a heat exchanger to recover the waste heat.

10.3.3 Steam turbines

Steam turbines are often used as the prime mover in larger CHP installations. Steam turbines can employ open or closed cycles, depending on whether or not the steam itself is used as the site heating medium. In the closed system, high-pressure steam from a boiler is forced through a turbine, which in turn rotates a generator. Heat is then recovered from the steam by passing it through a condenser on its way back to the boiler. In open cycle systems the steam exiting the turbine is used directly to meet site energy needs. The power produced by the steam turbine is therefore dependent on the extent to which the steam pressure is reduced through the turbine. The simplest open cycle arrangement is the *back-pressure* system, which employs a pressure regulator after the turbine, so that the steam is exhausted at the pressure required by the site. As the exhaust steam pressure is raised, so the temperature and heat output increase. However this increase in heat output is at the expense of the power output, which reduces. By regulating the exhaust steam pressure it is possible to control the heat to power ratio of the CHP plant thus creating a very flexible system. Lower steam pressures can be used in the summer when less heat is required, resulting in higher electricity generating efficiencies. In winter when higher temperatures are required, steam pressure can be raised. Consequently, the heat to power ratio of steam turbine CHP schemes can be variable, ranging from 3:1 to as much as 10:1 [4].

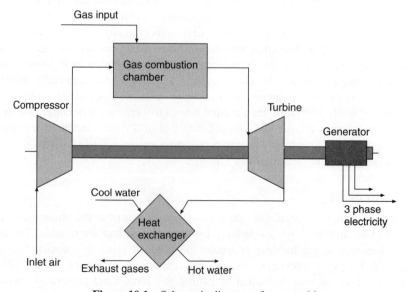

Figure 10.1   Schematic diagram of a gas turbine

They are therefore best suited to schemes in which there is a high all-year-round heat requirement. A typical *back-pressure* steam turbine CHP arrangement is shown in Figure 10.2.

Because a boiler is employed to produce the steam to drive a turbine, a wide variety of fuels can
be used, including refuse. Therefore steam turbines are a good solution for *waste-to-energy* CHP
schemes in which refuse is incinerated and the heat used to produce steam. In Scandinavia it is
common practice to burn the waste products from the timber industry to produce steam in CHP
schemes.

10.4 Micro-CHP systems

Stand-alone micro-CHP units are a popular solution for many small- and medium-sized commercial applications. Micro-CHP units use an internal combustion engine as a prime mover, and
generally comprise an engine, an electricity generator, a heat recovery system, an exhaust, and a
control system (as shown in Figure 10.3).

In a micro-CHP system, optimum efficiency is achieved by maximizing the heat recovered
from the engine and exhaust gases. In theory as much as 90% of the heat produced by the generation process can be recovered. Achieving this level of heat recovery requires the use of several heat exchangers, which makes the capital cost high. It is therefore more typical to recover
around 50 per cent of the fuel input as useful high-grade heat, with a further 10% recovered as
low-grade heat [3]. The high-grade heat can be used to provide heating water in the region of
70–90 °C and the low-grade heat to provide water at 30–40 °C [3]. Most of the heat is recovered
from the engine jacket, which has a temperature of approximately 120 °C, while the rest is
recovered from the exhaust gases, which can be at 650 °C [4]. Both sensible and latent heat can
be recovered from the exhaust gases.

Although most micro-CHP units provide low temperature hot water (LTHW) in the region
70–80 °C, it is equally possible to provide medium temperature hot water (MTHW) (i.e. 90–120 °C).

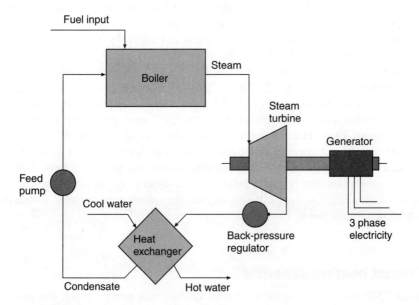

Figure 10.2   Schematic diagram of *back-pressure* steam turbine CHP system

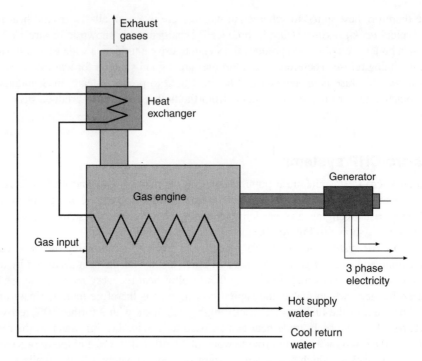

Figure 10.3   Schematic diagram of a micro-CHP unit

However, because of the higher water temperatures involved, heat recovery is reduced. Conversely, it is possible to increase heat recovery and therefore the efficiency of a micro-CHP system, by reducing the hot water supply temperature to below 70 °C. As most micro-CHP systems are required to produce domestic hot water (DHW), which must be stored at above 60 °C to prevent the growth of *Legionella* spp., in practice the flow water temperature should be 70 °C or above.

Micro-CHP units are often used in conjunction with boilers. In such systems the CHP unit should satisfy the base heating load, with the boilers only being used during periods of peak demand. This necessitates coupling the micro-CHP unit to the boilers, so that the two can work effectively together. In existing installations, where a CHP unit is replacing some old boilers, it is common practice to connect the CHP unit and boilers in series as this causes minimum interference to existing systems. In new installations, CHP units are often connected in parallel with boilers. Figure 10.4 illustrates both arrangements. No matter which arrangement, it is essential that the CHP unit operates as the lead 'boiler', as this maximizes its operating hours.

Many micro-CHP units incorporate a continuous monitoring facility as part of their control system. This enables building heat and power requirements to be monitored so that optimum performance of the plant is achieved. It also enables the system to be audited, so that the return on the capital investment can be calculated.

10.5 District heating schemes

Many larger CHP units are coupled to district heating schemes in which the pipework and pumping costs are dominant. In such schemes it is important to minimize both pipe diameters

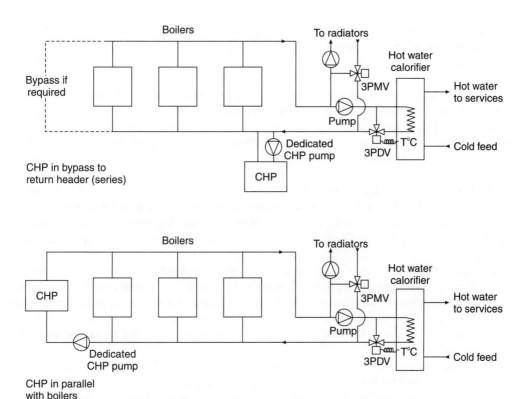

Figure 10.4   CHP piping arrangement. Crown copyright is reproduced with the permission of the
Controller of Her Majesty's Stationery Office and the Queen's Printer for Scotland [3]

and water flow rates, by operating at a peak flow water temperature of approximately 120 °C,
with a return water temperature of 70 °C. This reduces both capital and operating costs. It is also
common practice to vary supply water temperature with ambient air temperature so that system
heat losses are minimized.

If LTHW is required in individual buildings on a district heating scheme, this can be achieved
by installing remote heat exchangers in each building. This maintains hydraulic separation
between the district heating water and the LTHW and makes the overall system safe and
flexible. In Europe, variable temperature district heating schemes with heat exchangers are very
popular.

10.6 CHP applications

CHP systems are considered to be efficient users of primary energy because the waste heat
produced by the generation process is utilized to satisfy heating requirements. If 'waste' heat
cannot be utilized effectively, overall efficiency will drop dramatically. In simple terms, there is
no point in installing a CHP system in an application which does not have an all-year-round
demand for heat. CHP systems are therefore suitable for buildings such as leisure centres, swimming pools, hotels, hospitals, and residential establishments, all of which have extensive DHW
requirements for most or all of the year. Office buildings are generally thought to be unsuitable,
since they frequently have a cooling load for much of the year and are only open during the day

time. If however the heat from a CHP unit is used to drive absorption refrigeration plant, then CHP can become a feasible option for office buildings.

Although the operational costs associated with CHP systems are low, the capital costs are high. It is therefore desirable to run a CHP unit for as long as possible in order to achieve the greatest return on the initial capital investment. It has been calculated that in order to achieve a simple payback of 3 to 4 years it is necessary to operate a CHP unit between 4500 and 6000 hours per year [3], which is equivalent to approximately 12.3 to 16.5 hours of operation for each day of the year. It is much better to undersize a CHP unit than to oversize it, since this will ensure that the unit runs continuously when in operation, with any shortfall in output being made up by back-up boilers and bought-in electricity. It is therefore common practice to use the CHP unit to satisfy base heat load requirements. Ideally a CHP unit should be able to supply the entire summer heat load and a proportion of the winter load. Although it may be relatively small (possibly with a rated output of only 33–50% of the peak heating demand) it is possible to supply 60–90% of a building's annual heat requirement with a CHP unit because it supplies the base heat load.

In certain situations, where a CHP unit generates more electricity than can be consumed on site, it is possible to export power to the local utility company. This depends on the willingness of the utility company to purchase the electricity. It also requires the installation of an export meter. Therefore, for small-scale CHP installations it is not generally considered economic to export electricity. Micro-CHP units should therefore be sized so as not to exceed the base electrical load.

It is possible to use a CHP unit as a standby generator if so required. If used in this way its size will be governed by the required peak emergency electrical load. For normal operation it will be necessary to modulate down the output to match the reduced heat and power requirements, with the result that efficiency will be compromised. In such circumstances it may be more economical to install two smaller CHP units.

10.7 Operating and capital costs

The capital and installation costs of CHP plant can be significantly higher than those for conventional boiler plant. One significant cost which can easily be overlooked is the requirement of CHP systems to be synchronized in parallel with the local utility company's distribution grid, so that the grid and the CHP unit can work together to meet peak site electrical demand. This involves the installation of expensive electrical switching equipment. In contrast to the capital costs, the operating costs associated with CHP are relatively low and comprise the fuel and maintenance costs. For micro-CHP units maintenance costs are generally in the range of 0.5–2.0 p per kWhe of electricity generated [3], with the maintenance cost reducing for larger systems. Typical capital and maintenance costs for various sized CHP units are shown in Table 10.1.

10.8 CHP plant sizing strategies

In order to correctly size a CHP installation it is important to obtain as much accurate energy data as possible for the given application. Ideally these data should include:

- Monthly electricity and heat energy consumption data in kWh.
- Base and peak-load demands (in kW) for both electricity and heat.
- The operational characteristics of the particular application.
- Unit cost data for electricity and gas (or oil).

Table 10.1   CHP installation and maintenance costs (1996 data) [5]

CHP engine size (kWe)	Installed capital cost (£/kWe)	Maintenance cost (p/kWhe)
45	1230	1.04
54	1170	1.02
90	1020	0.98
110	960	0.95
167	810	0.89
210	730	0.85
300	660	0.79
384	605	0.73
600	520	0.62

Because it is important not to oversize a CHP plant it is advisable to undertake all the possible no-cost and low-cost energy efficiency measures before sizing the plant. This will avoid the CHP unit being oversized and should reduce the capital cost of the installation.

When determining the size of a CHP unit the most commonly used approach is to size the unit to meet the base heating load, as shown in Figure 10.5. This ensures that the CHP unit can run all year round, thus guaranteeing that the payback period on the initial capital investment is short. Back-up boilers can then be used to meet the peak-load heating requirements. A CHP unit sized in this way usually generates less electricity than is required to meet the base electrical demand and therefore additional electrical energy must be purchased all-year-round from the local utility company. An alternative approach is to size the CHP unit to meet the electrical base load. This usually means that for part of the year heat will have to be dumped because the heat produced by the CHP unit will exceed the base-load requirement. However, despite the dumping of heat this can be the most economic solution, since the unit cost of electricity can be as much as five times that of a unit of heat.

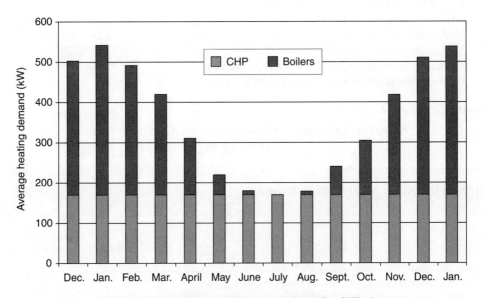

Figure 10.5   Base heat load sizing strategy for CHP plant

10.9 The economics of CHP

For most small-scale CHP applications three factors dominate economic viability. These are:

- The capital cost of the installation.
- The potential number of operating hours per year.
- The relative costs of 'bought-in' electricity and gas (or fuel oil).

If any of these three variables are not favourable, then a particular CHP scheme may become non-viable. Given that fuel prices can be unstable, the last point is of particular importance. If for example, the unit cost of mains electricity should fall or the cost of gas rise, there will come a point when a particular CHP unit ceases to be economically viable. Other lesser factors which may influence the economic performance of a CHP scheme are:

- The heat to power ratio of the particular CHP plant;
- The difference in maintenance costs between a CHP scheme and a conventional scheme;
- The cost of having mains electricity as a back-up system in case of breakdown or maintenance.

Given these costs, it is important to undertake a full economic appraisal of any proposed CHP scheme. Example 10.2 illustrates how a simple appraisal might be undertaken.

Example 10.2

An existing hotel building has an average electrical demand of 80 kWe and an average combined heating and hot water demand of 180 kW. The average annual load factor for the building is 0.75. The heating and hot water demand is currently served by two gas fired boilers and mains electricity is bought-in. It is proposed to install a micro-CHP plant which will run on gas and have a heat to power ratio of 1.7:1. The existing boilers will supplement the heat output from the CHP unit. If the initial cost of the CHP installation is £76 000, determine the simple payback period.

Data:

> Efficiency of existing boilers = 70%
> CHP unit electric power output = 80 kWe
> CHP unit gas power input = 286 kW
> Unit price of electricity = 5.0 p/kWh
> Unit price of gas = 0.9 p/kWh
> Existing plant maintenance cost = £1000 per year
> CHP scheme maintenance cost = £5000 per year

Solution

$$\text{Annual operating hours} = \text{load factor} \times \text{total hours per year}$$
$$= 0.75 \times 8760 = 6570 \text{ hours}$$

Considering the present scheme:

$$\text{Electricity cost} = \frac{80 \times 6570 \times 5.0}{100} = £26\,280.00$$

$$\text{Gas cost} = \frac{180 \times 6570 \times 0.9}{0.7 \times 100} = £15\,204.86$$

and

$$\text{Maintenance cost} = £1000.00$$

therefore

$$\text{Annual cost} = 26\,280.00 + 15\,204.86 + 1000.00$$
$$= £\,42\,484.86$$

Considering the proposed CHP scheme:

The average short-fall in CHP heat production $= 180 - (80 \times 1.7) = 44$ kW.

Therefore

$$\text{Annual CHP unit fuel cost} = \frac{286 \times 6570 \times 0.9}{100} = £16\,911.18$$

$$\text{Annual boiler fuel cost} = \frac{44 \times 6570 \times 0.9}{0.7 \times 100} = £3716.74$$

and

$$\text{Maintenance cost} = £5000.00$$

therefore

$$\text{Annual operating cost} = 16\,911.18 + 3716.74 + 5000.00 = £25\,627.92.$$

Now

$$\text{Pay back} = \frac{\text{Capital cost}}{\text{Annual cost saving}}.$$

Therefore

$$\text{Pay back} = \frac{76\,000}{(42\,484.86 - 25\,627.92)}$$
$$= 4.51 \text{ years}$$

While the analysis undertaken in Example 10.2 gives some indication of the economic viability of a CHP scheme, the method used is simplistic and has a number of inherent weaknesses. It assumes that the electrical and heating demands are constant at 80 kWe and 180 kW respectively. In reality this will not be the case. For long periods during the year demand will be higher than this, while at other times it will be lower. This means that during periods of high electrical demand (i.e. when the electrical demand exceeds 80 kWe), electricity will have to be purchased from the local utility company. However during periods of low demand the CHP unit will be producing electricity and heat which cannot be utilized. As a result the analysis overestimates the potential cost savings achievable through using CHP.

A more sophisticated approach which overcomes some of the short-falls described above is illustrated in Example 10.3.

Example 10.3

A new sports centre is to be built which will have a predicted annual heat load of 2600 000 kWh and an annual electrical load of 830 000 kWhe. The peak winter heating and hot water demand is predicted to be 1000 kW and the base heat demand is 350 kW. The base electrical demand is 130 kWe. The sports centre plant will operate for 5130 hours per year. Given the following data appraise the financial viability of three proposed schemes:

(a) conventional scheme in which boilers produce all the heat and electricity is purchased from a utility company.
(b) A CHP scheme in which the CHP plant is sized to meet the base electrical load.
(c) A CHP scheme in which the CHP plant is sized to meet the base heat load.

Data:

Efficiency of boilers = 70%
Mechanical efficiency of CHP unit = 30%
Efficiency of CHP electricity generator = 95%
Heat recovery efficiency of CHP unit = 70%
Unit cost of gas = 0.9 p/kWh
Unit cost of electricity = 5.0 p/kWh
Cost of maintaining boilers = 0.1 p/kWh
CHP scheme maintenance cost = 0.9 p/kWhe
Capital cost of boiler only scheme = £26.50 per kW
Capital cost of CHP scheme = £900 per kWe

Solution

(a) Considering the conventional scheme:

$$\text{Electricity cost} = \frac{830\ 000 \times 5.0}{100} = £41\ 500.00$$

$$\text{Gas cost} = \frac{2600\ 000 \times 0.9}{0.7 \times 100} = £33\ 428.57$$

and

$$\text{Maintenance cost} = \frac{2600\ 000 \times 0.1}{100} = £2600.00$$

Therefore

$$\text{Annual cost} = 41\ 500.00 + 33\ 428.57 + 2600.00 = £77\ 528.57$$

and

$$\text{Capital cost} = 26.50 \times 1000 = £26\ 500.00.$$

(b) Considering the CHP scheme, sized to meet the base electrical load:

$$\text{Fuel power input to CHP unit} = \frac{130}{0.3 \times 0.95} = 456.14 \text{ kW}$$

The waste heat produced by the CHP unit is passed through a heat exchanger with an efficiency of 70%, therefore:

$$\text{Recoverable heat power} = (456.14 \times (1 - 0.3)) \times 0.70 = 223.51 \text{ kW.}$$

Therefore

$$\text{Annual electricity produced by CHP unit} = 130 \times 5130 = 666\ 900 \text{ kWhe}$$

and

$$\text{Annual heat produced by CHP unit} = 223.51 \times 5130 = 1146\ 606.3 \text{ kWh}$$

Therefore

$$\text{Annual CHP unit fuel cost} = \frac{454.14 \times 5130 \times 0.9}{100} = £21\ 059.98$$

$$\text{Annual boiler fuel cost} = \frac{(2600\ 000 - 1146\ 606.3) \times 0.9}{0.7 \times 100} = £18\ 686.49$$

$$\text{Annual cost of electricity purchased} = \frac{(830\ 000 - 666\ 900) \times 5.0}{100} = £8155.00$$

and

$$\text{Maintenance cost} = \frac{0.9 \times 666\ 900}{100} = £6002.10.$$

Therefore

$$\text{Annual operating cost} = 21\ 059.98 + 18\ 686.49 + 8\ 155.00 + 6002.10$$
$$= £53\ 903.57$$

and

$$\text{Capital cost of CHP scheme} = 900.00 \times 130 = £117\ 000.00$$

Therefore

$$\text{Increased capital expenditure compared with Scheme (a)} = 117\ 000.00 - 26\ 500.00$$
$$= £90\ 500.00$$

and

$$\text{Annual operating cost saving} = 77\ 528.57 - 53\ 903.57 = £23\ 625.00$$
$$\text{(compared with Scheme (a))}$$

therefore

$$\text{Pay back on increased Capital expenditure} = \frac{90\ 500.00}{23\ 625.00} = 3.8 \text{ years}$$

(c) *Considering the CHP scheme, sized to meet the base heat load:*

$$\text{Heat produced for each kWe of electrical power generated} = \frac{223.51}{130} = 1.719$$

Therefore, the heat to power ratio of the CHP unit is 1.719:1. Assuming that the CHP unit is sized to meet the base heat load of 350 kW, then:

$$\text{Electrical power output from CHP unit} = \frac{350}{1.719} = 203.61 \text{ kW}$$

Unfortunately since the average electrical demand of the building is only 161.79 kWe, the CHP unit produces more electricity than can be consumed by the building. Unless the electricity can be exported to the local utility company, the CHP unit will either have to be reduced in size, or else its output will have to be modulated down considerably.

If it is assumed that electricity can be exported at, say, 3.0 p/kWhe, then:

$$\text{Annual revenue generated through exporting electricity} = \frac{(203.61 - 161.79) \times 5130 \times 3.0}{100}$$
$$= £6436.10$$

and

$$\text{Annual heat produced by CHP unit} = 350 \times 5130 = 1795\ 500 \text{ kWh}$$
$$\text{Annual electricity produced by CHP unit} = 161.79 \times 5130 = 830\ 000 \text{ k Whe}$$

NB: The CHP unit provides all the electricity for the building.

$$\text{Fuel power input to CHP unit} = \frac{203.61}{0.3 \times 0.95} = 714.42 \text{ kW.}$$

Therefore

$$\text{Annual CHP unit fuel cost} = \frac{714.42 \times 5130 \times 0.9}{100} = £32\ 984.77$$

$$\text{Annual boiler fuel cost} = \frac{(2600\ 000 - 1795\ 500) \times 0.9}{0.7 \times 100} = £10\ 343.57$$

and

$$\text{Maintenance cost} = \frac{0.9 \times (203.61 \times 5130)}{100} = £9400.67$$

Therefore

$$\text{Annual operating cost} = 32\ 984.77 + 10\ 343.57 + 9400.67 - 6436.10$$
$$= £46\ 292.91$$

and

$$\text{Capital cost of CHP scheme} = 900.00 \times 203.61 = £183\ 249.00$$

Therefore

$$\text{Increased capital expenditure (compared with Scheme (a))} = 183\ 249.00 - 26\ 500.00$$
$$= £156\ 749.00$$

and

$$\text{Annual operating cost saving (compared with Scheme (a))} = 77\ 528.57 - 46\ 292.91$$
$$= £31\ 235.66$$

Therefore

$$\text{Pay back on increased Capital expenditure} = \frac{156\ 749.00}{31\ 235.66} = 5.02 \text{ years}$$

Example 10.3 clearly demonstrates that both CHP schemes achieve substantial cost savings compared with the conventional scheme (a). However, it should be noted that although scheme (c), sized to meet the base heat load, produces the greatest annual cost savings, scheme (b) appears to be the more cost effective of the two proposals. This is because:

- The capital cost of scheme (c) is much higher than that of scheme (b); and
- Much of the electricity produced under scheme (c) is under-utilized (i.e. exported for a relatively low return).

Example 10.3 therefore reinforces the conclusion that it is unwise to oversize a CHP plant and confirms that it is preferable to size the CHP plant to meet the electrical base load.

References

1. Beggs, C. B. (1996). A method for estimating the time-of-day carbon dioxide emissions per kWh of delivered electrical energy in England and Wales. *Building Services Engineering Research and Technology*, **17**, (Number 3), 127–134

2. Department of the Environment, Transport and the Regions (2001). Environmental Reporting – Guidelines for Company Reporting on Greenhouse Gas Emissions, www.detr.gov.uk/ environment/envrp/gas/ 05.htm, 8 May.
3. Department of the Environment (1989). Guidance notes for the implementation of small scale packaged combined heat and power. Good Practice Guide 1.
4. Department of the Environment (1992). Introduction to large-scale combined heat and power. Good Practice Guide 43.
5. Williams J., Griffiths, T. and Knight, I. (1996). Sizing chp for new hospitals. *Building Services Journal*, November, pp. 41–3.

Bibliography

Cost model: Combined heat and power system analysis (1997). *Building Services Journal*, January, pp. 17–19.

Department of the Environment (1989). Guidance notes for the implementation of small scale packaged combined heat and power. Good Practice Guide 1.

Department of the Environment (1992). Introduction to large-scale combined heat and power. Good Practice Guide 43.

Eastop, T. D. and Croft, D. R. (1990). *Energy efficiency for engineers and technologists*, (Chapter 8). Longman Scientific & Technical.

Williams, J., Griffiths, T. and Knight, I. (1996). Sizing chp for new hospitals. *Building Services Journal*, November, pp. 41–3.

11

Energy efficient air conditioning and mechanical ventilation

Much energy is wasted in buildings through the use of inappropriate air conditioning and mechanical ventilation systems. This situation has arisen because building designers are often ignorant of the issues associated with air conditioning and also because designers of air conditioning systems are more interested in minimizing first costs rather than reducing overall energy consumption. There are, however, a number of new and innovative technologies, which have the potential to reduce energy consumption greatly. This chapter discusses the issues associated with the design of air conditioning and mechanical ventilation systems, and introduces some of these new low energy technologies.

11.1 The impact of air conditioning

Over the last 40 years or so, there has been a trend towards large deep plan buildings with highly insulated envelopes. This trend, coupled with the increased use of personal computers and the use of high illumination levels, has meant that many buildings over-heat and thus require cooling for large parts of the year, even in countries which experience cool temperate climates. Over the years boilers have steadily reduced in size and the use of air conditioning and mechanical ventilation has increased. When it is also considered that most of the Earth's population live in countries which have warm or hot climates, it is not difficult to appreciate that the provision of adequate cooling and ventilation is a much greater global issue than the provision of adequate heating. Unfortunately, many building designers are not fully aware of this simple fact, with the result that a great number of poorly designed buildings are erected, relying on large air conditioning systems in order to maintain a tolerable internal environment.

The contribution of mechanical cooling towards overall global energy consumption should not be underestimated. In the UK alone, it has been estimated that approximately 10 000 GWh of electrical energy is consumed per annum by air conditioning equipment [1]. This represents approximately 14% of all the electrical energy consumed in the commercial and public administration sectors in the UK. Of this figure, approximately 5853 GWh is consumed solely by refrigeration plant, the rest being consumed by fans, pumps and controls [2]. In the USA, the energy consumed by air conditioning equipment is much higher. Indeed, in many of the southern states in the USA, electrical demand increases by 30–40 % during the summer months solely due to the use of air conditioning equipment [3]. As a result of this, the utility companies in the southern states of the USA have to install excess generating capacity to meet the summer peak, even though for most of the year this plant remains idle, which is clearly a very uneconomical situation. The problems faced by the electrical utilities in the USA are typical of many companies operating in warm climates throughout the world. In some countries, electrical demand is so high during the summer months that the authorities ration electricity by restricting the capacity of power cables which enter properties. In doing so they force building owners and users to utilize low energy design solutions.

It is a common misconception that most of the energy consumed by air conditioning plant is associated with the operation of refrigeration machines. This is not the case. In reality much more energy is consumed by air handling plant. A recent study of typical 'standard' air conditioned office buildings in the UK found that refrigeration plant consumed 13% of total electricity consumption, while fans, pumps and controls consumed 26.5% of all the electrical energy consumed. In this type of office building approximately 35% of the total energy costs were spent on running the air conditioning and mechanical ventilation systems [4]. A summary of the results of this study is presented in Table 11.1.

The environmental impact of air conditioning equipment is considerable. Air conditioning is uniquely catastrophic from an environmental point of view, since it:

- Can contribute directly to atmospheric ozone depletion through the leakage of harmful refrigerants.
- Contributes directly to global warming through the leakage of refrigerants which are powerful greenhouse gases.
- Contributes to global warming by consuming large amounts of electricity and indirectly releases large quantities of CO_2 into the atmosphere.

A full discussion of the environmental problems associated with refrigerants is beyond the scope of this book. Nevertheless, a brief discussion of environmental matters is perhaps relevant here. Until recently, both CFCs and HCFCs were extensively used as refrigerants. Although CFCs and HCFCs are known to be potent greenhouse gases, they are much more infamous for being potent ozone depletors. Indeed, it was the serious threat to the ozone layer which ended production of CFCs in 1995 under the Montreal Protocol [5]. Since then there has been heavy reliance on the use of HCFC 22 as an alternative to CFCs. While HCFC 22 is far more ozone friendly than CFC 11 or 12, it is still a potent greenhouse gas, having a GWP of 1700 [6]. However, under the Montreal Protocol, HCFCs are also being phased-out, with production due to cease completely by 2030 [7]. Consequently, the chemical manufacturers are currently developing a new generation of refrigerants, HFCs, to replace the old CFCs and HCFCs. Unfortunately, while HFCs are ozone benign they are still strong greenhouse gases. Notwithstanding this, the relative effect which refrigerants have on global warming is often overestimated. The contribution to global warming made by escaping refrigerants is far outweighed by the indirect CO_2 emissions resulting from the electrical consumption of refrigeration machines. This is graphically illustrated by Figure 11.1 which shows the relative contribution to global warming of associated CO_2 emissions compared with that of a variety of refrigerants [8].

It can be seen from Figure 11.1 that the indirect contribution of air conditioning equipment towards global warming is considerable. It has been estimated that in the UK alone, 4.2 million tonnes of CO_2 per annum are directly attributable to the use of air conditioning refrigeration plant [2]. It is therefore not surprising that governments around the world are putting pressure on building designers to reduce or eliminate the need for mechanical cooling.

Table 11.1 Energy consumption in various UK office buildings [4]

	Naturally ventilated cellular		Naturally ventilated open-plan		Air conditioned standard		Air conditioned prestige	
	Good practice (kWh/m²)	Typical (kWh/m²)	Good practice (kWh/m²)	Typical (kWh/m²)	Good practice (kWh/m²)	Typical (kWh/m²)	Good practice (kWh/m²)	Typical (kWh/m²)
Heating and hot water (gas or oil)	79	151	79	151	97	178	107	201
Mechanical cooling	0	0	1	2	14	31	21	41
Fans, pumps and controls	2	6	4	8	30	60	36	67
Humidification	0	0	0	0	8	18	12	23
Lighting	14	23	22	38	27	54	29	60
Office equipment	12	18	20	27	23	31	23	32
Catering (gas)	0	0	0	0	0	0	7	9
Catering (electricity)	2	3	3	5	5	6	13	15
Other electricity	3	4	4	5	7	8	13	15
Computer room (where applicable)	0	0	0	0	14	18	87	105
Total gas or oil	79	151	79	151	97	178	114	210
Total electricity	33	54	54	85	128	226	234	358

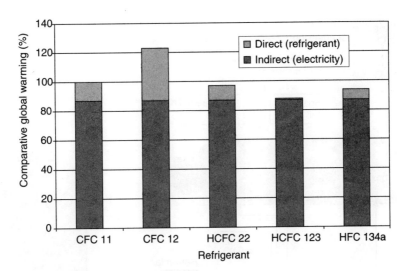

Figure 11.1 Comparison of the direct and indirect contribution of various refrigerant machine types towards global warming[8]

11.2 Air conditioning systems

This chapter is not intended to be a text on the fundamentals of air conditioning design, but rather a discussion of the application of air conditioning in buildings. Before discussing in detail the issues which affect the energy consumption of air conditioning systems, it is necessary first to describe briefly the nature and operation of a generic air conditioning installation. It should be noted that in this text, for ease of reference, the term *air conditioning* is used in its loosest sense to describe any system which employs refrigeration to cool air in buildings.

Figure 11.2 shows a simple air conditioning system which illustrates many generic features. The system employs an air handling unit (AHU) to blow air at a constant volume flow rate through ducts to a room space. Stale air is then removed from the room space via an extract duct using a return fan. In order to save energy, it is common practice to recirculate a large proportion (e.g. 70%) of the return air stream using mixing dampers located in the AHU. It is also common practice to propel air along the ducts at velocities in excess of 5 m/s. This ensures that duct-work sizes are kept to a minimum. Room space temperature is controlled by varying the temperature of the incoming supply air; in winter, air is supplied at a temperature higher than that of the room space, while in summer the air is supplied at a temperature lower than that of the room space. In this way a comfortable environment can be maintained all-year-round in the room space.

The AHU in Figure 11.2 comprises:

- A mixing damper section to mix the incoming fresh air with recirculated air;
- A filter to clean the air;
- A heating coil (usually a hot water coil fed from a boiler but sometimes electric);
- A cooling coil to cool and dehumidify the air;
- A reheat coil to accurately control the air temperature and to compensate for any over-cooling by the cooling coil; and
- A centrifugal fan to draw the air through the AHU and to push it through the ductwork.

In the case of the AHU shown in Figure 11.2, a direct expansion (DX) refrigeration coil is used to cool the supply air. This coil is the evaporator of a refrigeration system, and contains liquid

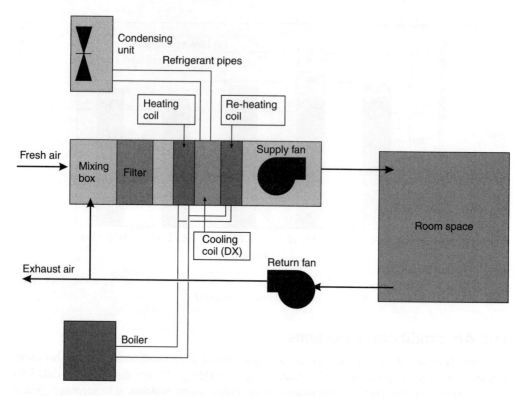

Figure 11.2   A simple ducted air conditioning system with a direct expansion (DX) cooling coil

refrigerant which boils at a low temperature and pressure (e.g. at 5 °C and 584 kPa) to become a low pressure vapour. As liquid boils it draws large quantities of heat from the air stream and thus cools it. At the other end of the refrigerant pipes to the DX coil, a condensing unit is located which comprises a compressor, a condenser and a fan. The heat taken from the supply air stream by the DX coil is rejected at the condenser to the atmosphere.

The system shown in Figure 11.2 is generic and is typical of many systems found throughout the world. There are however some variations which are worthy of note. In many applications in hot countries there is no requirement for heating, and so the heating coils are removed, leaving only the DX cooling coil. Similarly there may be no requirement to supply fresh air, in which case the filter and mixing dampers can be omitted. Common examples of such simple systems are the 'through the wall unit' (see Figure 11.3) and the 'split unit' systems (see Figure 11.4). These systems are inexpensive and easy to maintain, and not surprisingly, are very popular in many hot countries.

One of the major disadvantages of the systems described above is that in larger installations they require many condensing units to be placed on the outside of buildings. This can be both unsightly and impractical. So in many larger buildings, a superior solution is to install a central-ized refrigeration machine, known as a chiller, to produce chilled water (e.g. at 7 °C) which can then be pumped to a number of remote AHUs (see Figure 11.5). In this type of system, each AHU is fitted with a chilled water cooling coil instead of a DX coil. Chilled water cooling coils are superior to DX coils, because they facilitate closer control of the supply air temperature. Centralized chillers also have an environmental advantage over remote DX systems, insomuch

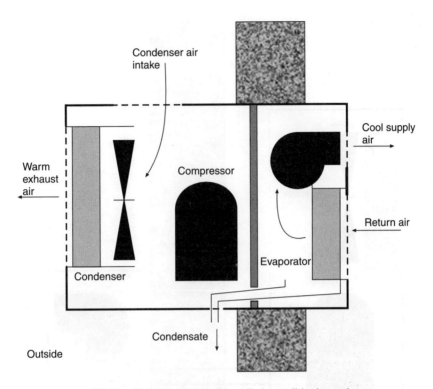

Condenser air
intake

Cool supply
air

Warm
exhaust
air

Compressor

Condenser

Evaporator

Return air

Outside

Condensate

Figure 11.3   A through the wall air conditioning unit

as there are fewer refrigeration circuits to maintain, resulting in lower risk of refrigerant leaks. Chillers can utilize either air or water cooled condensers. Water cooled condensers are more efficient than the air cooled variety, but usually require a cooling tower, and are therefore a potential *Legionella pneumophila* hazard. For this reason, air cooled chillers have become more popular than water cooled chillers, because despite being less efficient they present no health hazard.

11.3 Refrigeration systems

Most air conditioning plant relies on some form of vapour compression refrigeration machine to remove heat from air. Figure 11.6 shows a schematic diagram of a simple, single-stage, vapour compressor refrigeration system, similar to that found in many air conditioning systems.

The vapour compression refrigeration cycle operates as follows:

1 Low pressure liquid refrigerant in the evaporator boils to produce low pressure vapour. The heat required to boil and vaporize the liquid within the evaporator is taken from an air or water stream passing over the outside of the evaporator.
2 After leaving the evaporator, the low pressure refrigerant vapour enters the compressor where both its temperature and its pressure are raised by isentropic compression.
3 The high pressure refrigerant vapour then passes to the condenser where it is cooled and liquefied. The heat extracted in the condenser is released to the environment either directly by forcing air over the outside of the condenser, or indirectly using a secondary fluid, usually water, and a cooling tower.

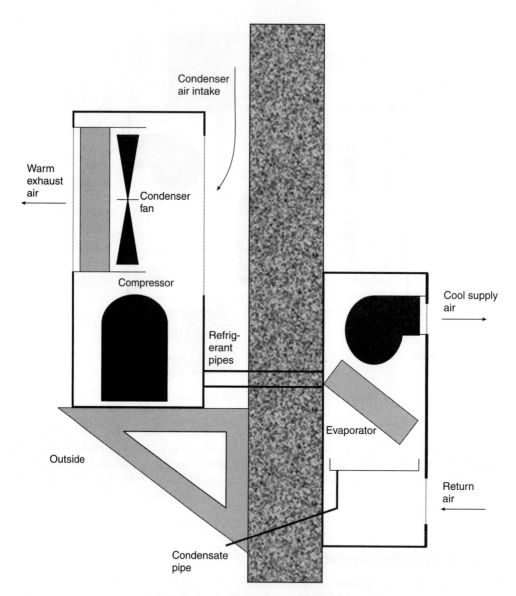

Figure 11.4 A split unit air conditioning system

4   The high pressure liquid refrigerant then passes from the condenser to the expansion valve, where its pressure is lowered, and approximately 10% of the liquid 'flashes' (i.e. instantly turns from a liquid to a vapour) thus cooling the remainder of the liquid. The cooled low pressure liquid then flows into the evaporator and the cycle begins all over again.

The boiling point of refrigerants varies with pressure. At low pressures, refrigerants boil at low temperatures (e.g. 2 °C), while at much higher pressures the boiling point of the refrigerant is significantly raised (e.g. 35 °C). In this way refrigerants can be vaporized and condensed at different temperatures, simply by altering system pressure. Since condensing and evaporating

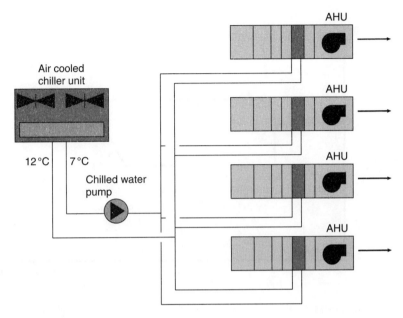

Figure 11.5   An air cooled chiller system with multiple air handling units

temperatures correspond to particular pressures, they are normally 'measured' using pressure gauges located before and after the compressor.

Figure 11.7 shows a plot of the vapour compression cycle on a pressure/enthalpy diagram. The refrigeration capacity of the system is the amount of cooling which the plant can achieve and is proportional to the length of the line between points 4 and 1. The power input to the system is through the compressor drive, and is represented by the line 1 to 2. Line 2 to 3 represents the heat rejection at the condenser. Line 3 to 4 represents the passage of the refrigerant through the expansion device and is a constant enthalpy process. It should be noted that the heat rejected by the condenser is equal to the total energy input at the evaporator and at the compressor. Note also that as the condensing pressure decreases, so too does the power input from the compressor.

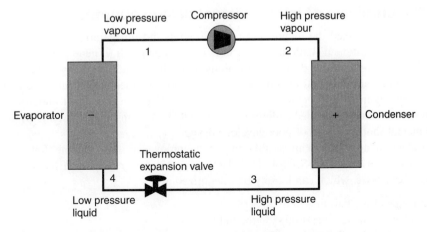

Figure 11.6   Vapour compression refrigeration cycles

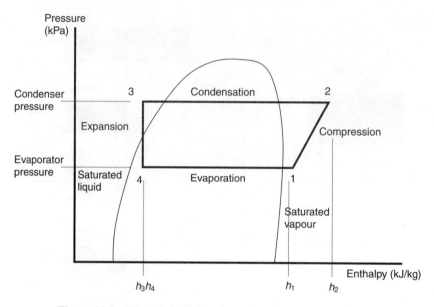

Figure 11.7   Pressure/enthalpy chart of vapour compression process

The overall 'efficiency' of a vapour compression machine is normally described by the COP. The COP of a refrigeration machine is the ratio of the refrigeration capacity to the power input at the compressor. It can be expressed as (referring to Figure 11.7):

$$COP_{ref} = \frac{h_1 - h_4}{h_2 - h_1} \tag{11.1}$$

where h is the specific enthalpy of refrigerant (kJ/kg).

The higher the COP, the more efficient the refrigeration process. In the UK, refrigeration machines generally exhibit COPs in the range of 2.0–3.0 [2].

11.4 The problems of the traditional design approach

Having briefly discussed the nature of air conditioning and refrigeration systems, the overall design of 'air conditioned' buildings must now be considered. It is generally the case that in buildings the air conditioning design is something of an afterthought. In many buildings the form and the envelope are designed in complete isolation from the mechanical services. Usually, air conditioning engineers are required to design and install systems which fit unobtrusively into buildings (i.e. behind suspended ceilings etc.); often these systems are required to overcome the environmental short-comings of poor envelope design.

The traditional approach to air conditioning is to employ a constant volume flow rate system, similar to that described in Section 11.2. However, this design approach has a great many inherent weaknesses, which can loosely be categorized as:

- Weaknesses of the building design;
- Weaknesses of the refrigeration system; and
- Weaknesses of the air system.

11.4.1 Building design weaknesses

Buildings have to function properly in a great many harsh environments around the world. In hot desert climates they are required to keep their occupants cool, while in polar regions keeping warm is the important issue. Buildings should therefore be designed so that the external envelope is the primary climate modifier, with the internal mechanical services simply fine tuning the shortcomings of the envelope. This may seem an obvious statement, but clearly it is not one which is fully understood by many building designers. In Texas, which has a hot arid climate, there are many glass clad office buildings. It would be intolerable to work in these buildings were it not for the use of very large air conditioning systems to compensate for the inappropriate building envelope. This apparently ludicrous situation comes about for three specific reasons:

1 Energy efficiency is often a low priority; minimizing the first cost is usually the prime consideration.
2 Design professionals often work in isolation from each other, and have little understanding of building physics, or of how buildings function when occupied.
3 There is great incentive to maintain the status quo. Building services design engineers are usually paid a fee which is a fixed proportion of the total capital cost of the building services. Consequently, there is little incentive to reduce the capacity of the mechanical building services.

In order to avoid the creation of energy wasteful buildings, it is important that energy efficiency be at the forefront of the designer's mind. Critical decisions made at the design stage have huge ramifications on both capital and operating costs. If the envelope is a poor climate modifier then the building will experience high summer heat gains and high winter heat losses, necessitating the installation of large boilers and refrigeration chillers. These items of equipment may, however, only operate at peak load for a few hours per year, with the result that for most of the year they operate very inefficiently at part load. Conversely, if the building envelope success-fully attenuates the winter and summertime peaks, then the plant sizes can be greatly reduced, resulting in the plant operating near its rated capacity for a much greater part of the year. Clearly, the latter situation is a much better utilization of capital expenditure than the former.

11.4.2 Refrigeration system weaknesses

The strategy of installing refrigeration plant to meet the peak summertime cooling load of build-ings often results in greatly oversized mechanical plant which operates inefficiently, at part load, for most of the year. It can also result in a greatly oversized electrical installation, since larger cables, transformers and switchgear must be installed to meet the peak refrigeration capacity. Not only does a system such as this have a high capital cost, it is also expensive to run since it uses peak time electricity. In hot countries it may also incur high electrical demand related charges. Refrigeration chillers are often oversized because:

- System designers overestimate peak building heat gains to ensure that plant is not undersized.
- System designers make design assumptions which are widely inaccurate. For example, in most buildings the actual cooling load is much less than the design cooling load. This discrepancy primarily occurs because designers assume very high 'office equipment' heat gains, which in practice rarely materialize.
- Refrigeration chillers are often rated for hot climates such as that found in the USA. So when these machines are installed in a temperate climate location such as in the UK, their condensers are oversized and so they operate at part load even when under peak load conditions.

The general oversizing of refrigeration machines results in very poor overall energy efficiencies. It has been estimated that of the refrigeration plant currently in operation in the UK, the average air cooled chiller has a working gross COP of approximately 1.9, while water cooled chillers exhibit an average gross COP of 3.0 [2]. These low figures are mainly due to the design of most of the refrigeration chillers used in the UK, namely machines which use thermostatic expansion valves and maintain a relatively fixed condensing pressure under part load conditions. These machines display poor COPs under part load conditions, which is unfortunate since for most of the year they operate in this state.

There are a number of alternative design strategies which can be used dramatically to reduce the size of refrigeration plant and improve operating costs. These are:

- The use of a solar defensive building envelope, incorporating features such as external shading and solar reflective glass to reduce the peak cooling load.
- The use of a thermally massive structure to absorb both internal and solar heat gains during peak periods.
- The use of night ventilation to purge the building structure of heat accumulated during the daytime.
- The use of ice thermal storage to shift some of the peak time cooling load to the nighttime.
- The use of a floating internal air temperature strategy, which allows internal temperatures to rise when conditions are exceptionally hot.

11.4.3 Air system weaknesses

The strategy of using an *all air* system to condition room air has the major disadvantage that it necessitates the transportation of large volumes of air and is thus inherently inefficient. It is generally the case that much larger volumes of supply air are required to sensibly cool room spaces than are required for the purpose of pure ventilation. Consequently, the provision of sensible cooling using an *all air* system results in large fans and associated air handling equipment, and also in large ceiling (or floor) voids to accommodate oversized duct work. Although an expensive solution, resulting in increased energy and capital expenditure, *all air* systems are still very popular, despite the existence of superior alternatives which use chilled water to perform the sensible cooling. One such alternative strategy is the use of chilled ceilings to perform sensible cooling, while reserving the ductwork system for ventilation purposes only. This strategy results in greatly reduced fan and ductwork sizes.

Another major draw back of the constant volume approach, described in Section 11.2, is that air duct and fan sizes are determined by the peak summertime condition which may only last for a few hours per year. For the rest of the year the fans push large volumes of air around needlessly, with the result that energy consumption on air handling is large. Many air conditioning system designers argue that constant volume systems have the potential to provide large amounts of free cooling during the spring and autumn seasons. This unfortunately is a misconception since oversized fans consume such large quantities of electrical energy that any saving in refrigeration energy is wiped out. The evidence for this can be seen in Table 11.1, where in the air conditioned buildings almost twice as much electrical energy is consumed by air distribution systems compared with the refrigeration machines. One alternative strategy which overcomes this problem is to adopt a variable air volume system in which the quantity of the air handled reduces with the cooling load.

It is common practice in ducted air systems to size the main ducts assuming air velocities of 4 m/s to 7 m/s. Designers use these relatively high air velocities in order to keep duct sizes to a minimum. Unfortunately, the use of such high air velocities results in large system resistances.

The fan power consumed in a ducted air system can be determined using eqn (11.2). Fan power,

$$W = \dot{v} \times \Delta P_{total} \tag{11.2}$$

where $\dot{v}$ is the volume flow rate of air discharged by the fan (m³/s), and ΔP_{total} is the total system pressure drop or resistance (Pa).

From eqn (11.2) it can be seen that there is a linear relationship between fan power and system resistance; the higher the system resistance, the higher the energy consumed by the fan. The system resistance is the sum of the system static pressure drop and the velocity pressure drop. The velocity pressure in particular strongly influences the pressure drop across ductwork bends and fittings. From eqn (11.3) it can be seen that the pressure drop across a ductwork fitting is a function of the square of the air velocity.

Pressure drop across ductwork fitting:

$$\Delta P = k \times (0.5 \, \rho v^2) \tag{11.3}$$

where k is the velocity pressure loss factor for fitting, ρ is the density of air (kg/m³), v is the velocity of air (m/s), and $(0.5 \, \rho v^2)$ is the velocity pressure (Pa).

Given eqns (11.2) and (11.3) it is not difficult to see that the use of high air velocities (i.e. in the region of 5 m/s) results in high fan powers and high energy consumption. However, if air velocities are reduced to approximately 1 m/s to 2 m/s, as is the case in some *low energy* buildings, then fan energy consumption falls dramatically.

11.5 Alternative approaches

The critique of the traditional approach to the design of air conditioning systems presented in Section 11.4 highlights its many shortcomings and hints at a number of possible solutions. There are several alternative low energy strategies which may be employed to overcome the disadvantages of the conventional approach. Although inter-linked, for ease of reference these alternative strategies can loosely be categorized as follows:

- Using passive solar defensive and natural ventilation measures to reduce the need for air conditioning;
- Splitting the sensible cooling and ventilation roles into two separate but complementary systems;
- Using low velocity and variable air volume flow systems;
- Using the thermal mass of buildings to absorb heat which can then be purged by a variety of ventilation techniques;
- Using thermal storage techniques to shift the peak cooling load to the nighttime;
- Using displacement ventilation techniques;
- Using evaporative cooling; and
- Using desiccant cooling techniques.

Many of the energy conservation techniques listed above are discussed in detail in this chapter. Some of the techniques which relate specifically to building envelope design are specifically dealt with in Chapter 13.

11.6 Energy efficient refrigeration

Although a number of alternative cooling strategies are discussed in this chapter, there are still many applications which demand the use of conventional refrigeration plant. It is therefore necessary to understand the factors which influence the energy consumption of conventional vapour compression refrigeration machines. The major factors influencing energy performance are:

- The evaporating and condensing temperatures used;
- The type of refrigerant used;
- The type of compressor and condenser used;
- The defrost method used on the evaporator; and
- The system controls.

Each of these factors can have a profound effect on overall energy consumption and are therefore worthy of further investigation.

11.6.1 Evaporators

The efficiency of vapour compression systems increases as the evaporating temperature increases. Generally, the higher the evaporating temperature used, the greater the system COP and the lower the energy consumption. It has been estimated that a rise in the evaporating temperature of 1 °C results in an operating cost reduction of between 2% and 4% [9]. It is therefore desirable to maintain the evaporating temperatures as high as is practically possible. Maximum heat transfer across the evaporator should be achieved in order to prevent the evaporating temperature from dropping. In practice, this can be achieved by increasing the fluid flow across the evaporator, or by increasing its surface area. Also, in order to ensure high evaporating temperatures it is essential that good control of the system be maintained.

On air cooling applications where the evaporator may be operating below 0 °C the fin spacing must allow for ice build-up. In order to maintain an adequate airflow through the evaporator it is necessary to defrost the coil periodically. Defrosting techniques involve either the use of an electric heating element built in to the coil or periodically reversing the refrigeration cycle so that the evaporator effectively becomes a hot condenser. While essential for the correct operation of the system, the defrost process can be a potential source of energy wastage. It is therefore important that the defrost operation only be initiated when absolutely necessary and that the defrost heat be evenly distributed over the whole of the fin block. Prolonged defrosting is energy wasteful and therefore the defrost cycle should be stopped as soon as possible. If not controlled and monitored properly defrost systems can needlessly waste large amounts of energy.

11.6.2 Condensers

Condensing temperature can have a dramatic influence on system COP, with lower condensing temperatures usually resulting in lower operating costs. It is estimated that a 1 °C drop in condensing temperature reduces operating costs by approximately 2–4% [9]. However, if the condensing pressure fluctuates widely, problems can occur on machines which utilize thermostatic expansion valves. This is because such valves are unable to reliably control refrigerant flow at low pressure differentials. In order to overcome this problem, these machines often employ some form of condenser pressure control to raise the condenser pressure artificially. This results in unnecessarily high energy consumption, which could be avoided if electronic expansion devices were used instead.

Jo Coleman

Information Update Service

Butterworth-Heinemann

FREEPOST SCE 5435

Oxford

Oxon

OX2 8BR

UK

Keep up-to-date with the latest books in your field

Visit our website and register now for our FREE e-mail update service, or join our mailing list and enter our monthly prize draw to win £100 worth of books. Just complete the form below and return it to us now! (FREEPOST if you are based in the UK)

www.bh.com

Please Complete In Block Capitals

Title of book you have purchased:..

....

Subject area of interest:..

Name:...

Job title:...

Business sector (if relevant):...

Street:..

Town:... County:...

Country:.. Postcode:.....................................

Email:...

Telephone:..

How would you prefer to be contacted: Post ☐   e-mail ☐   Both ☐

Signature:... Date:...

☐ Please arrange for me to be kept informed of other books and information services on this and related subjects (✔ box if not required). This information is being collected on behalf of Reed Elsevier plc group and may be used to supply information about products by companies within the group.

FOR OFFICE USE ONLY

Butterworth-Heinemann,
a division of Reed Educational
& Professional Publishing Limited.
Registered office: 25 Victoria Street,
London SW1H 0EX.
Registered in England 3099304.
VAT number GB: 663 3472 30.

**BUTTERWORTH
HEINEMANN**

A member of the Reed Elsevier plc group

There are three basic condenser systems commonly in use: air-cooled condensers, water-cooled condensers and evaporative condensers, each of which has its own peculiarities. Air-cooled condensers are by far the most popular heat rejection system. They generally comprise a fin and tube heat exchanger in which refrigerant vapour condenses. Air is forced over the heat exchanger by fans. Well-designed condensers should operate at a temperature no higher than 14 °C above the ambient air temperature [10]. In larger air cooled systems, condenser pressure is often controlled by switching off or slowing down fans. Although this practice is inefficient it does save on energy consumed by the condenser fans. One important advantage of air-cooled condensers is that they present no *Legionella pneumophila* risk.

Water-cooled condensers are much more compact than their air-cooled counterparts and comprise a shell containing refrigerant, through which pass water-filled tubes. Secondary cooling water flows through these tubes to a cooling tower where the heat is finally rejected through an evaporative cooling process. In an efficient system, the temperature rise of the water passing through the condenser should be 5 °C, with a difference of 5 °C existing between the condensing temperature and the temperature of the water leaving the condenser [10].

Water-cooled systems are considerably more efficient than air cooled systems, with the former requiring a cooling tower air flow rate of approximately 0.04 m³/s to 0.08 m³/s per kW of rejected heat and the latter requiring 0.14 m³/s to 0.2 m³/s to perform the same task [11]. However, there is a risk of *Legionella pneumophila* bacteria breeding in cooling towers, if they are not monitored and regularly treated with biocides. For this reason water-cooled condensers have become less popular in recent years.

Evaporative condensers are less popular than either air or water cooled condensers. They comprise refrigerant condensing tubes which are externally wet and over which air is forced. Evaporation of the water on the outside of the tubes increases the heat rejection rate and so this type of condenser is more efficient than its air-cooled counterpart. Evaporative condensers do, however, pose a *Legionella pneumophila* risk.

11.6.3 Compressors

The compressor is the only part of a vapour compression system which consumes energy. It is therefore important that the factors which influence compressor performance are well understood. The efficiency of a compressor can be expressed in several ways. In terms of overall energy consumption the most critical 'efficiency' is isentropic efficiency, which is defined as follows:

$$\text{Isentropic efficiency} = \frac{\text{ideal 'no loss' power input}}{\text{actual power input to shaft}} \times 100 \qquad (11.4)$$

It should be noted that isentropic efficiency does not take into account motor and drive inefficiencies, which must be allowed for when determining the overall efficiency of a compressor. With most types of compressor, particularly screw and centrifugal compressors, efficiencies fall dramatically under part-load operation. Compressor motor efficiency also decreases at part-load. In general therefore, part-load operation should be avoided if high efficiencies are to be maintained.

Part-load operation is the main reason for poor refrigeration plant efficiency. Many refrigeration machines spend less than 20% of the year operating at their nominal design condition. During the rest of the year they operate at part-load, partly because of cooler ambient temperatures and partly because of reduced cooling duties. Unless these part-load conditions are properly allowed for at the design stage, it is likely that overall system COP will be poor. It is therefore important to select a compressor which exhibits good part-load efficiency.

Multi-cylinder reciprocating compressors achieve reasonable part-load efficiencies because they are able to unload cylinders so that output is reduced in steps (e.g. 75, 50 and 25%). Variable speed drives can also be used. These give good flexible control and can achieve reasonable efficiencies above 30% of full-load [12].

11.6.4 Expansion devices

In a refrigeration machine the expansion valve is used to reduce the pressure of the returning liquid refrigerant from the condensing pressure to the evaporating pressure. It also controls the flow of liquid refrigerant to the evaporator. It is therefore important that expansion valves be correctly selected and installed, since incorrect operation of the expansion valve can lead to reduced energy efficiency.

Thermostatic expansion valves are the most commonly used type of refrigerant regulation device. They regulate the flow of refrigerant through the system by opening and closing a small orifice using a 'needle' connected to a diaphragm (as shown in Figure 11.8). The diaphragm responds to pressure changes created inside a control phial which senses the temperature of the refrigerant leaving the evaporator. Both the phial and the valve contain refrigerant. As the load on the evaporator changes, so the temperature of the refrigerant leaving the evaporator also changes. The control phial senses these changes in temperature and automatically adjusts the refrigerant flow to accommodate the load changes.

The major disadvantage of thermostatic valves is that they cannot cope with large pressure differentials, such as those created when the condensing pressure is allowed to float with ambient

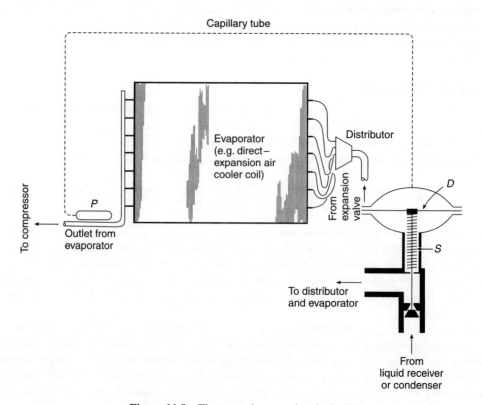

Figure 11.8   Thermostatic expansion device[11]

air temperature. Thermostatic expansion valves tend to operate unsatisfactorily at less than 50% of their rated capacity [13]. Therefore, in refrigerating machines using thermostatic expansion valves it is often necessary to maintain an artificially high condensing temperature during conditions of low ambient temperature.

It can be seen from the discussion in Section 11.6.2 that if the condensing pressure is allowed to fall, the COP increases. Therefore in theory, system efficiency should improve when ambient air temperature falls. Unfortunately, due to the operational characteristics of thermostatic expansion valves it is not possible to take advantage of this situation. Thermostatic expansion valves are therefore inherently inefficient. In recent years an alternative technology has arisen which overcomes this problem. By using electronic expansion valves it is possible to allow condensing pressures to drop while still maintaining a constant evaporating pressure. Unlike conventional thermostatic expansion valves, which operate on the degree of superheating in the evaporator, electronic expansion valves employ a microprocessor which constantly monitors the position of the valve, the temperature of the liquid in the evaporator, and the temperature of the vapour leaving the evaporator. They can therefore respond quickly to fluctuations in load and are not dependent on large differential pressures between the condenser and the evaporator. Consequently, it is possible under low ambient conditions to allow the condensing pressure to fall and the COP to improve.

11.6.5 Heat recovery

Vapour compression refrigeration machines are simply a specific form of heat pump. This means that they reject large quantities of waste heat at the condenser. In many applications, through a little careful thought at the design stage, it is possible to utilize this waste heat profitably to reduce energy costs. In refrigeration installations heat can be recovered from:

- The compressor discharge gas, which is generally in the region 70–90 °C, but can be as high as 150 °C in some installations; and
- The condenser, which is generally 10–30 °C above ambient temperature.

High quality waste heat can be recovered from the compressor discharge gas by using a desuperheater heat exchanger. This device recovers heat from the high temperature vapour before it reaches the condenser. At this point in the cycle the vapour is at its highest temperature and therefore the heat transfer is at its greatest. It is important to locate desuperheaters above condensers so that if any refrigerant vapour condenses, the liquid can safely drain away.

Heat may also be recovered at the condenser. However, because condensing temperatures should ideally be as low as possible, any heat recovered here will inevitably be at a relatively low temperature. The quality of this heat should however, be sufficient to pre-heat domestic hot water. If higher condensing temperatures are envisaged as a result of any proposed heat recovery scheme, careful analysis should be undertaken to ensure that economic benefit will accrue. Remember, there are many so called 'energy saving' schemes in existence which have actually increased energy costs!

11.7 Splitting sensible cooling and ventilation

When considering air conditioning, it is possible to save substantial amounts of energy by splitting up the sensible cooling and ventilation roles into two separate systems. This also enables fan and duct sizes to be greatly reduced. One commonly used method of separating these two roles is to use a fan coil system, which circulates chilled water through water/air heat exchangers (incorporating recirculation fans) located in room mounted units. Although fan coils function well in many applications, they can take up valuable room space and also be noisy.

In addition, they utilize recirculation fans which consume energy. A novel alternative approach is the use of passive chilled ceilings or beams, which comprise a cold surface mounted at high level within a room space. Chilled ceilings and beams perform room sensible cooling and leave the ducted air system to perform the ventilation and latent cooling roles.

Chilled ceilings have a very slim profile and usually comprise a metal pipe coil bonded to a flat metal plate (as shown in Figure 11.9). They can be fixed directly to the soffit of a structural floor slab, thus eliminating the need for an expensive suspended ceiling. They are usually designed to have an average surface temperature of approximately 17 °C, which can be achieved in practice by supplying low grade chilled water at about 13 °C. Because relatively high water temperatures are used it means that chiller evaporating temperatures can be high, resulting in very good refrigeration COPs being achieved.

The cooling power of chilled ceilings varies with the individual design used and the extent to which air turbulence occurs across the heat exchanger surface. In general an output of approximately 50 W/m^2 is considered to be the maximum that can be obtained. This means that chilled ceilings are not suitable for applications which experience very high internal heat gains. Heat is transferred to the chilled ceiling by natural convection and radiation. Approximately 50% of the heat transfer occurs when warm air at the top of the room space comes into contact with the cool surface. The remaining heat transfer is by radiation from room occupants and other warm surfaces within the room space.

The radiative cooling capability of chilled ceilings is of particular importance and is worthy of further comment. At relative humidities below 70%, the thermal comfort of building occupants is primarily governed by room air temperature and room surface temperature. In order to assess and quantify relative thermal comfort, a number of thermal comfort indices have been developed. Although these comfort indices vary slightly from each other, they all seek to quantify the convective and radiative heat transfer to and from an occupant within a room space. The most widely used thermal comfort index in the UK is *dry resultant temperature*. Provided

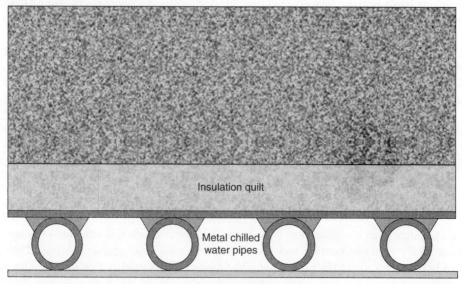

Insulation quilt

Metal chilled
water pipes

Metal ceiling tile

Figure 11.9 Typical chilled ceiling

room air velocities are less than 0.1 m/s (which is usually the case), *dry resultant temperature* can be expressed as:

$$t_{res} = 0.5t_a + 0.5t_r \qquad (11.5)$$

where t_r is the mean radiant temperature (°C), and t_a is the air temperature (°C).

Mean radiant temperature is the average surface temperature of all the surfaces 'seen' in a room space. It can be either measured indirectly using a globe thermometer or calculated from surface temperatures. For most cuboid shaped rooms, the mean radiant temperature in the centre of the room can be expressed as:

$$t_r = \frac{\Sigma(A_s \cdot t_s)}{\Sigma A_s} \qquad (11.6)$$

where A_s is the area of each component surface (m²), and t_s is the temperature of each component surface (°C).

By installing a large surface area of chilled ceiling at 17 °C, it is possible to substantially reduce room *mean radiant temperature*. Consequently, air temperatures can be allowed to rise to say, 23 °C or 24 °C without any deterioration in perceived comfort, with the result that energy can be saved.

A variation on the chilled ceiling theme is the passive chilled beam system (see Figure 11.10). Passive chilled beams work on a similar principle to chilled ceilings, but they achieve a much greater cooling output (e.g. 185 W/m²) and exhibit a much higher convective cooling component than chilled ceilings (e.g. approximately 85%). However, this can cause problems, since uncomfortable downdraughts can be created.

11.7.1 Ventilation

By utilizing chilled ceilings or beams, it is possible to free-up ductwork systems to concentrate solely upon the ventilation and latent cooling tasks. Normally the fresh air requirement of room occupants is in the region 8 l/s to 12 l/s per person. However, when using chilled ceilings or beams

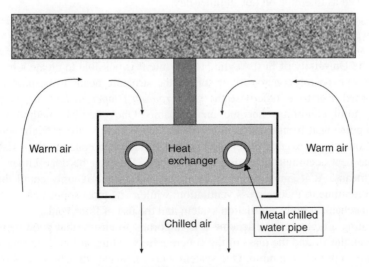

Figure 11.10   Typical chilled beam

the ducted air ventilation system also has to perform all of the room latent cooling. In order to do this it is common for the ventilation rate to be increased slightly to 18 l/s per person [14].

With chilled ceilings and beams it is important to ensure that the room air moisture content be maintained at a low level, otherwise condensation may occur on the cool surface and ultimately 'internal rain' may be formed. This makes it important to ensure that the incoming ventilation air is dehumidified, so that the room dew point temperature remains a few degrees below the surface temperature of the chilled ceilings. An ideal air condition for a room incorporating a chilled ceiling is 24 °C and 40% relative humidity.

11.8 Fabric thermal storage

Although, a full discussion of the role of fabric thermal storage is contained in Chapter 13, a few words on the subject are perhaps relevant here. The widespread use of suspended ceilings and carpets in buildings means that otherwise thermally heavyweight structures are converted into thermally lightweight ones. These low admittance buildings are not able to absorb much heat and so surface temperatures tend to rise, with the result that there is a great need to get rid of the heat gains as they occur. This is one of the main reasons why air conditioning has become such an essential requirement of so many office buildings. By contrast, if the mass of the building structure is exposed, then a high admittance environment is formed, and the thermal capacity of the structure can be utilized successfully to combat overheating.

The creation of a high admittance environment by exposing thermal mass has implications on the comfort of occupants. It can be seen from the discussion in Section 11.7, that it is the *dry resultant temperature* and not the air temperature which is critical when establishing a comfortable environment. By exposing the 'mass' of a building it is possible to reduce the *mean radiant temperature* within the space, and thus the *dry resultant temperature*. So if an office building with openable windows and exposed concrete floor soffits has a mean radiant temperature of, for example, 20 °C and the air temperature in the space is 28 °C, then the perceived temperature (i.e. the *dry resultant temperature*) in the space will be only 24 °C. While an internal air temperature of 28 °C is generally considered unacceptable, a dry resultant temperature of 24 °C will be perceived as tolerable on hot summer days.

In buildings which employ 'thermal mass' to control internal temperatures, it is common practice to expose concrete floor soffits to create a high admittance environment, as can be seen in examples such as the Queens Building at De Montfort University [15] and the Elizabeth Fry Building at the University of East Anglia [16]. While it is possible to create a high admittance environment by exposing concrete floor soffits, the structure needs to be purged periodically of heat absorbed over time, otherwise the *mean radiant temperature* of the room spaces will steadily rise until conditions become unacceptable. One effective method which can be employed to purge heat from the structure of buildings is night venting. Night venting involves passing cool outside air over or under the exposed surface of a concrete floor slab so that it is purged of the heat accumulated during the day time. This can be done either by natural or mechanical means. At its most rudimentary night venting may simply entail the opening of windows at nighttime to induce cross ventilation, while at its most sophisticated it may involve a dedicated mechanical night ventilation system and the use of floor voids.

When creating a night venting scheme it is important to ensure that good thermal coupling occurs between the air and the mass of the concrete floor, whilst at the same time ensuring that fan powers are kept to a minimum. One system which manages to achieve this is the Swedish Termodeck hollow concrete floor slab system (see Chapter 13). The Termodeck system has been used effectively in many locations, throughout northern Europe and in the UK [16,17] to produce buildings which are both thermally stable and energy efficient.

11.9 Ice thermal storage

In the exposed concrete soffit system described in Section 11.8, the cooling process is effectively 'load shifted' to the nighttime by using a passive thermal storage technique. While this can be very effective, the system is limited to cool temperate countries where nighttime temperatures are low enough to purge accumulated heat. In hotter climates this strategy is impossible and so an alternative approach to load shifting is required. One alternative technology is ice thermal storage, which utilizes low cost nighttime electricity to produce a 'cold store' for use during the daytime. The technique involves running refrigeration chillers during 'off-peak' hours to produce an ice store. During the daytime when electricity prices are high, the ice is melted to overcome building or process heat gains. The principal advantages of the system are as follows:

- Refrigeration energy costs can be significantly reduced, as a substantial portion of the cooling is undertaken using off-peak electricity.
- The capital cost of the refrigeration plant can be significantly reduced, if both the chillers and the store combine to satisfy the peak cooling load requirement.
- If an ice store is coupled with a conventional refrigeration plant, then it is possible to run the chiller constantly at 100% of its rated capacity and thus operate it in an efficient manner.
- If an ice store is coupled with an electronically controlled refrigeration plant, then it is possible to minimize the refrigeration energy expended. This is because the refrigeration plant will be running at nighttime when ambient temperatures are low and so operating COPs will be high.
- Any electricity maximum demand charges incurred by the system will be significantly lower than those incurred by conventional refrigeration plant.
- By installing additional ice stores it is possible to increase the overall capacity of existing air conditioning installations without purchasing new chillers or upgrading electrical systems.
- Ice storage systems enable CO_2 emissions to be reduced through load shifting [18] and can also reduce the quantity of refrigerant used.

Ice storage systems are generally associated with air conditioned commercial or public buildings. However, ice storage systems have also been used successfully in process industries which experience large and predictable cooling loads. In this type of application it is often the case that a relatively small refrigeration machine can, over a long period of time, generate a large ice store. The ice store can then be melted over a relatively short period of time, to satisfy the peak cooling load. In this way small refrigeration machines can be used to satisfy very large cooling loads, with the result large capital savings can be made on refrigeration plant. In addition, capital cost savings can be made on electrical cables and switch gear, which is of particular importance for applications in remote locations.

11.9.1 Control strategies

Ice thermal storage systems can be operated in a variety of ways, with the major control strategies being *full storage*, *partial storage* and *demand-limited storage*.

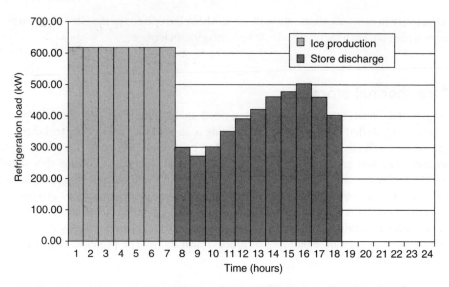

Figure 11.11   Full storage strategy

(a) *Full storage:* Under a *full storage* control strategy the total daytime cooling load is shifted to the nighttime, with the chillers producing an ice store during the period when off-peak electricity charges apply. During the daytime the ice store is discharged to meet the building or process cooling load, as shown in Figure 11.11. While being the most effective of all the control strategies in terms of energy costs, *full storage* has the major drawback that the ice store and chiller plant required are much larger than for the other control strategies. Due to its prohibitively high capital cost *full storage* is rarely used.

(b) *Partial storage: Partial storage* is the collective term given to those ice storage control strategies which require both the chiller plant and the ice store to operate together to satisfy the daytime cooling load. During periods in which the building or industrial process experiences a cooling load, the ice store and the chiller plant work simultaneously to satisfy the cooling load. The advantage of *partial storage* is that both the store and the chiller plant are substantially smaller than would be the case for a *full storage* installation and thus the capital cost is lower. This makes *partial storage* a very popular option. The umbrella term *partial storage* can be sub-divided into two separate and distinct sub-strategies, namely *chiller priority* and *store priority*.

Under a *chiller priority* control strategy the refrigeration plant runs continuously through both the ice production and the store discharge periods. During the daytime the refrigeration plant carries out the base-load cooling and the ice store is used to top-up the refrigeration capacity of the chiller plant (see Figure 11.12), which would otherwise be unable to cope with the peak demand.

Under a *chiller priority* strategy it is possible to achieve reductions in the region of 50 % in chiller capacity when compared with a conventional refrigeration installation. The capital cost of installing an ice store can therefore be off-set against the capital cost saving arising from the reduction in chiller capacity.

The philosophy behind the *store priority* control strategy is the opposite of the *chiller priority* strategy. Under a *store priority* strategy the ice store is given priority over the chiller during the daytime (see Figure 11.13). The objective of this strategy is to minimize the operation of

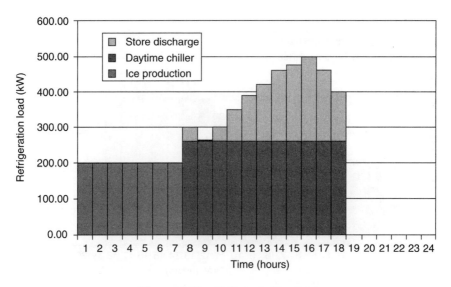

Figure 11.12 Chiller priority strategy

the refrigeration plant during periods when electricity prices are high. The refrigeration chiller is only used to top-up the refrigerating energy released by the ice store.

(c) *Demand-limited storage:* The object of a *demand-limited* control strategy is to limit peak electrical demand by shifting the cooling load out of periods in which the peak demand naturally occurs (see Figure 11.14). This greatly reduces the overall maximum demand of the installation and improves the overall load factor of the building, putting the operators in a stronger position when it comes to negotiating electricity supply contracts with the utility companies.

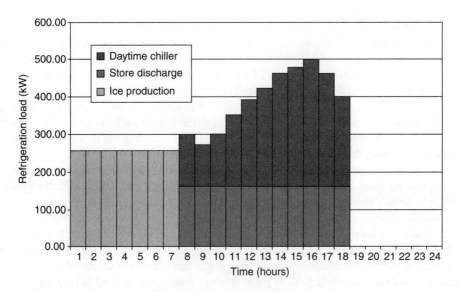

Figure 11.13 Store priority strategy

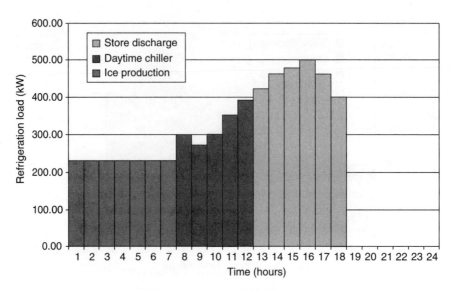

Figure 11.14   Demand limiting strategy

A *demand-limited* control strategy is particularly useful in situations where a utility company offers an electricity tariff which has either high unit charges or high demand charges for part of the daytime (e.g. from 12 am–6 pm), as is often the case in hot countries during the summertime. Under these circumstances, during the period for which peak charges apply, the cooling load should be entirely satisfied by the refrigeration energy released from the ice store.

11.9.2 Ice thermal storage systems

In broad terms ice storage systems fall into two main categories, static systems and dynamic systems. Static systems have the general characteristic that ice is melted in the same location as it is generated. Unlike static systems where the ice remains in one location throughout the entire operation of the installation, in dynamic systems the ice, once formed, is transported by some means to another location where it comes into contact with the working fluid, which is usually water.

The ice bank system shown in Figure 11.15 is typical of a static ice system. It consists of an insulated water storage tank, which contains a submerged bundle of small tubes. These tubes are evenly spaced within the tank volume, often in a spiral or serpentine form. During ice production a glycol/water solution at a sub-zero temperature is circulated through the tubes. This causes the water in the tank to freeze solid. During the discharge cycle, the ice is melted by the same glycol/water solution, this time circulating at a temperature above 0°C.

The most widely used dynamic system is the ice harvester. Ice harvesters have been used in the dairy industry for many years. They consist of an open insulated tank, above which a number of vertical refrigerant evaporator plates are located. Water is trickled over the surface of the plates so that it becomes frozen. Typically, within about 20 minutes an 8–10 mm thick layer of ice can be built up. The ice is harvested by removing it from the evaporator and allowing it to fall into the tank below. This process is achieved by interrupting the flow of liquid refrigerant through the evaporator plates and diverting hot discharge gas through them so that their surface temperature reaches approximately 5 °C. A photo-electric switch can be used to stop ice production when the ice in the sump reaches the required level. To discharge the ice store, system water is circulated through the ice sump. A typical ice harvester installation is shown in Figure 11.16.

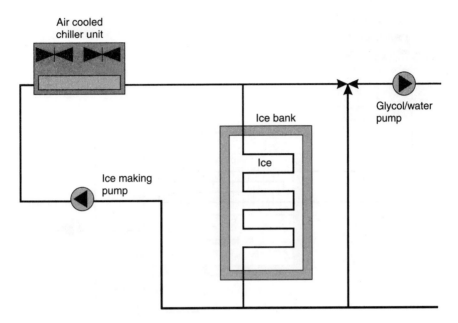

Figure 11.15   Ice bank system

11.9.3 Sizing of ice storage systems

The design calculations used to size ice storage systems depend on the precise control strategy which is adopted [19]. If a *chiller priority* control strategy is adopted then eqns (11.7–11.13) should be used. For a *store priority* strategy eqns (11.14–11.16) should be used.

$$Q_{st} + Q_{ch} = Q_j \tag{11.7}$$

where Q_{st} is the refrigeration energy contained within the ice store (kWh), Q_{ch} is the refrigeration energy produced by chiller plant when operating in the day time (kWh), and Q_j is the daily cooling load (energy) under design condition (kWh).

Under a *chiller priority* control strategy it is intended that the chiller plant should operate at full capacity throughout the day time period. However, it is not always possible to achieve this. A chiller plant will often operate at below its rated capacity for part of the day time. Consequently, eqn (11.7) must be modified to accommodate this:

$$Q_{st} + Q_{ch} = Q_j + Q_u \tag{11.8}$$

where Q_u is the unused chiller refrigeration energy (kWh).

The evaporating temperatures experienced by the refrigeration plant are much lower during the ice production than those experienced during day-time operation. Consequently, during the store charging period the chiller plant will experience reduced refrigerating capacity. It can therefore be stated that:

$$Q_{ch} = P_r \cdot H \tag{11.9}$$

and;

$$Q_{st} = P_r \cdot k_r \cdot h \tag{11.10}$$

where P_r is the rated duty of chiller under daytime operation (kW), k_r is the reduction factor for chiller producing ice, H is the duration of daytime chiller operation (hours), and h is the duration of ice production period (hours).

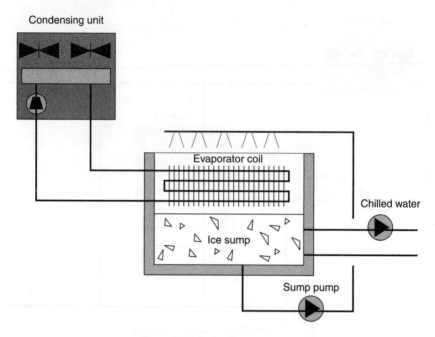

Figure 11.16   Ice harvester system

Therefore

$$Q_{st} + Q_{ch} = P_r \cdot (H + k_r \cdot h) \tag{11.11}$$

By combining eqns (11.8) and (11.11) it can be shown that:

$$P_r = \frac{Q_j + Q_u}{H + k_r \cdot h} \tag{11.12}$$

By combining eqns (11.8) and (11.9) it can be shown that:

$$Q_{st} = Q_j + Q_u - H \cdot P_r \tag{11.13}$$

In order to derive the plant sizing equations for a *store priority* control strategy, a slightly different approach is taken to that for the *chiller priority* equations. The concept of peak cooling load (P_m) is introduced. It can therefore be stated that:

$$Q_{st} + H \cdot P_r = H \cdot P_m - Q_v \tag{11.14}$$

where P_m is the peak cooling load experienced by building (kW), and Q_v is the unused ice storage capacity (kWh).

Therefore

$$Q_{st} = H \cdot P_m - Q_v - H \cdot P_r \tag{11.15}$$

By combining eqns (11.10) and (11.14) the following is produced:

$$P_r = \frac{H \cdot P_m - Q_v}{H + k_r \cdot h} \tag{11.16}$$

The process involved in sizing ice thermal storage systems is illustrated in Example 11.1.

Example 11.1

An office building has a peak daily cooling load of 5210 kWh, with a maximum instantaneous cooling duty of 620 kW. Given the following data:

(i) Determine the size of the ice store and chiller plant required for chiller priority, store priority and full storage control strategies, and for conventional chiller only system.
(ii) Determine the daily costs for the options outlined in (i) above; and
(iii) Determine the system capital costs for the options outlined in (i) above.

Data:

Off-peak electricity period = 00.00–07.00 hours
Peak electricity period = 07.00–24.00 hours
Air conditioning operation period = 08.00–18.00 hours
Day-time average COP = 3.00
Ice production COP = 2.25
Chiller capacity reduction = 0.75
factor for ice production
Peak unit charge = 5.50 p/kWh
Off-peak unit charge = 2.57 p/kWh
Capital cost of ice storage = 25 £/kWh
Capital cost of refrigeration chiller = 240 £/kW

NB: Assume that there is no unused refrigeration energy or ice store capacity in the process.

Solution

(i) *Conventional chiller only system:* There is no ice store so the chiller must have a refrigeration capacity of 620 kW.

Full storage control strategy: The chiller is not in operation during the daytime and so the ice store is required to satisfy all the daytime cooling load. Thus a large ice store is required to be built up over the 7 hour off-peak period. Therefore:

$$\text{Ice store capacity} = 5210 \text{ kWh}$$

and

$$\text{Nominal chiller duty} = \frac{5210}{7 \times 0.75} = 992.4 \text{ kW}$$

Chiller priority control strategy: The office air conditioning system is operational for 10 hours.
Therefore:

$$\text{Nominal chiller duty} = \frac{5210}{10 + (0.75 \times 7)} = 341.6 \text{ kW}$$

and

$$\text{Store capacity} = 5210 - (10 \times 341.6) = 1793.6 \text{ kWh}$$

Store priority control strategy: The peak summertime cooling duty is 620 kW. Therefore:

$$\text{Nominal chiller duty} = \frac{10 \times 620}{10 + (0.75 \times 7)} = 406.6 \text{ kW}$$

and

$$\text{Store capacity} = (10 \times 620) - (10 \times 406.6) = 2134.4 \text{ kWh}$$

(ii) *Conventional chiller only system*:

$$\text{Daily energy cost} = \frac{5210 \times 5.50}{3.0 \times 100} = £95.52$$

Full storage control strategy:

$$\text{Daily energy cost} = \frac{5210 \times 2.57}{2.25 \times 100} = £59.51$$

Chiller priority control strategy:

$$\text{Daily energy cost} = \frac{1793.6 \times 2.57}{2.25 \times 100} + \frac{(10 \times 341.6) \times 5.50}{3.0 \times 100}$$

$$= £83.11$$

Store priority control strategy:

$$\text{Daily energy cost} = \frac{2134.4 \times 2.57}{2.25 \times 100} + \frac{(5210 - 2134.4) \times 5.50}{3.0 \times 100}$$

$$= £80.77$$

(iii) *Conventional chiller only system*:

$$\text{Capital cost of installation} = 620 \times 240 = £148\ 800.00$$

Full storage control strategy:

$$\text{Capital cost of installation} = (992.4 \times 240) + (5210 \times 25) = £368\ 426.00$$

Chiller priority control strategy:

$$\text{Capital cost of installation} = (341.6 \times 240) + (1793.6 \times 25) = £126\ 824.00$$

Store priority control strategy:

$$\text{Capital cost of installation} = (406.6 \times 240) + (2134.4 \times 25) = £150\ 944.00$$

Results summary:

	Conventional chiller only system	Chiller priority	Store priority	Full storage
Chiller duty	620 kW	341.6 kW	406.6 kW	992.4 kW
Store capacity	n.a.	1793.6 kWh	2134.4 kWh	5210 kWh
Capital cost	£ 148 800	£ 126 824	£ 150 944	£ 368 426
Peak daily cost	£95.52	£ 83.11	£ 80.77	£ 59.51
Pay-back period*	n.a.	−4.85 years	0.40 years	16.71 years

* Indicative payback periods only, since calculated daily costs only apply to the peak summertime day.

From the results shown in Example 11.1 it can be seen that the initial choice of control strategy has a huge impact on the economic viability of any ice storage scheme. Given the inordinately large capital cost associated with *full storage* systems it is not surprising that these systems are rarely installed. The most popular strategy used is the *chiller priority* strategy which, as in the case in Example 11.1, can result in capital cost savings and a negative payback period.

11.10 Evaporative cooling

So far, most of the cooling techniques discussed in this chapter have utilized the vapour compression refrigeration cycle to perform air cooling. There are however a number of alternative technologies, such as evaporative cooling and desiccant cooling which can be utilized to perform air cooling. Direct evaporative cooling is perhaps the simplest of all the air cooling techniques and is extremely energy efficient. It relies on an adiabatic heat exchange between air and water in which the air is both sensibly cooled and humidified. Most direct evaporative cooling systems comprise an open porous matrix over which water is trickled and through which air can pass (see Figure 11.17). As air passes over the wetted media, water evaporates and so the air becomes more humid. In order to evaporate the water needs a 'package' of latent heat energy; this it takes from the air stream, with the result that the air is sensibly cooled. Figure 11.18 shows the evaporative cooling process on a psychrometric chart. It should be noted that the whole

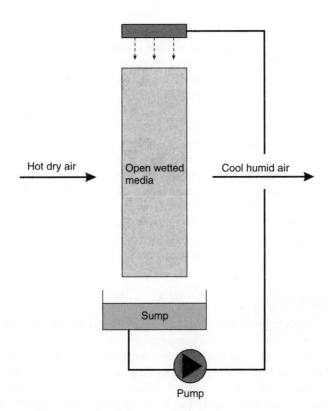

Figure 11.17   A direct evaporative cooler

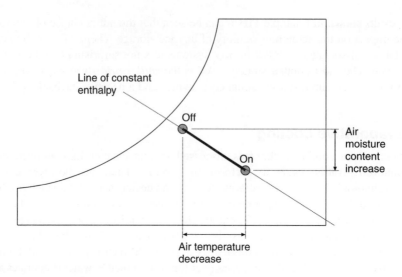

Figure 11.18   The direct evaporative cooling process

process is adiabatic and it follows the line of constant enthalpy on the psychrometric chart, which approximates to the line of constant wet-bulb temperature.

While direct evaporative coolers generally exhibit efficiencies of about 85 per cent [20], their sensible cooling effectiveness depends very much on the dryness of the air entering the cooler. If the air is very dry, then a large amount of sensible cooling will be achieved. Conversely, if the air has a high relative humidity, very little sensible cooling will be achieved. Not surprisingly, therefore, evaporative coolers have been used extensively for many years in hot arid countries, where they are often referred to as *desert coolers*. Direct evaporative cooling is very cost effective and eliminates the need for any environmentally unfriendly refrigerants.

One major disadvantage of direct evaporative cooling is that it greatly raises the relative humidity and moisture content of the air entering the room space, which may ultimately cause discomfort to room occupants. This problem can be overcome by introducing a heat exchanger to create an *indirect* evaporative cooling system. With indirect systems it is standard practice to place an evaporative cooler in the room exhaust air stream coupled to a flat plate heat recuperator. By using this arrangement the cool but humid exhaust air stream can be used to sensibly cool the incoming fresh air supply stream. It should be noted that there is no moisture exchange between the two air streams and so the supply air remains relatively dry. Figure 11.19 shows the indirect evaporative cooling process on a psychrometric chart. Indirect evaporative coolers will usually achieve an effectiveness of at least 60% and can achieve effectiveness ratings as high as 85% [20].

11.11 Desiccant cooling

Another alternative to the conventional vapour compression refrigeration cycle is to use a heat driven cycle. It has long been understood that desiccant materials such as silicon can be used to dehumidify air. Such systems pass moist air over surfaces which are coated with a desiccant substance. As the moist air passes across these surfaces the desiccant material absorbs moisture from the air, thus dehumidifying the air stream. In order to drive off the moisture absorbed by the desiccant surface, the desiccant has to be physically moved into a hot dry air stream. In the case of the desiccant wheel system (one of the most commonly used desiccant systems) the

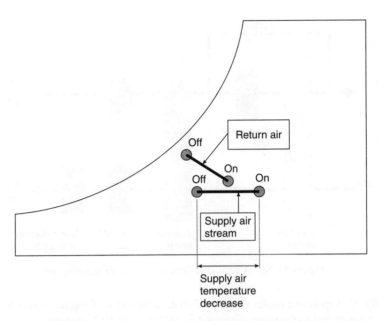

Figure 11.19   The indirect evaporative cooling process

moisture laden section of the wheel rotates slowly at approximately 16 revs/hour from the moist air stream to the hot dry air stream where it is regenerated.

Recently desiccant cooling systems have been developed which combine a desiccant wheel with a thermal wheel in a single AHU to produce a system which is capable of heating, cooling and dehumidifying air with little or no need for conventional refrigeration [21]. Such systems have the potential to reduce both energy costs and environmental pollution when compared with conventional refrigeration systems. From an environmental point of view the desiccant cooling system has the advantage that electrical energy consumption is replaced by heat consumption, which produces much less CO_2.

A typical desiccant cooling AHU is shown in Figure 11.20. It comprises a thermal wheel and a desiccant wheel located in series. On the supply side, after the thermal wheel, a supplementary cooling coil or an evaporative cooler may be located if so required. A heating coil may also be located after the thermal wheel for use in winter if required. An evaporative cooler is located in the return air stream before the thermal wheel so that the heat transfer across the thermal wheel is increased. The desiccant cycle is an open heat driven cycle; the 'driver' for the cycle is the regeneration heating coil located in the return air stream after the thermal wheel and before the desiccant wheel.

The psychrometric chart shown in Figure 11.21 illustrates the desiccant cooling and dehumidification process. During the summertime warm moist air at, for example 26 °C and 10.7 g/kg moisture content, is drawn through the desiccant wheel so that it comes off at say, 39 °C and 7.3 g/kg moisture content. The psychrometric process line for the air passing through the desiccant wheel on the supply side has a gradient approximately equal to that of a winter-time room ratio line of 0.6 on the psychrometric chart. The supply air stream then passes through the thermal wheel where it is sensibly cooled to say, 23 °C. The air then passes through a small DX or chilled water cooling coil where it is sensibly cooled to a supply condition of say, 17 °C and 7.3 g/kg moisture content. It should be noted that if humidity control is not required in the space, t;hen the cooling coil may

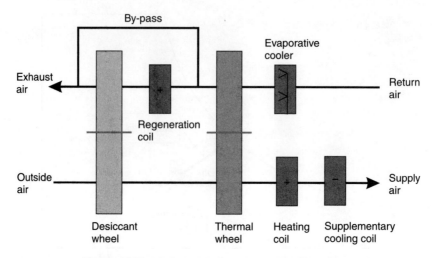

Figure 11.20   A typical desiccant cooling air handling unit

be replaced by an evaporative cooler with an adiabatic efficiency of approximately 85%, in which case air may be supplied to the room space at say, 16.2 °C and 10.2 g/kg moisture content.

On the return side, air from the room space at, for example, 22 °C and 8.6 g/kg moisture content passes through an evaporative cooler so that it enters the thermal wheel at approximately 16.7 °C and 10.8 g/kg moisture content. As the return air stream passes through the thermal wheel it is sensibly heated to approximately 33 °C. The air stream is then heated up to approximately 55 °C in order to regenerate the desiccant coil. It should be noted that in order to save energy approximately 20% of the return air stream bypasses the regenerating coil and the desiccant wheel.

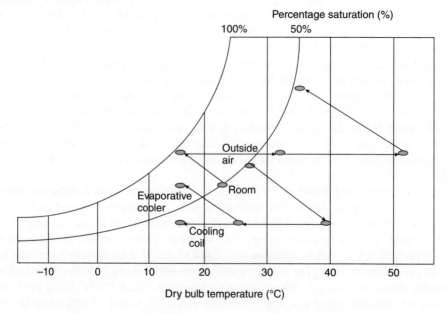

Figure 11.21   Desiccant system in cooling/dehumidification mode

During the wintertime much of the heat for the supply air stream comes from recovered heat from the thermal wheel. Although the desiccant wheel could in theory be used as an additional heat exchanger, in practice it is not particularly effective due to its low rotational speed, and is therefore not normally used. Should further sensible heating be required this can be achieved either by locating a heating coil in the supply air stream after the thermal wheel, or by using radiators within the room space. In addition, an evaporative cooler on the supply side may be utilized to humidify the incoming air stream if so required.

It has been shown that the use of desiccant cooling can result in energy cost savings ranging from 14% to 50% depending on the application and cooling load [14]. Surprisingly, unlike conventional refrigeration systems, operating costs are at their lowest when desiccant cooling systems are operating under part-load [14]. It is also worth noting that desiccant cooling systems are not well suited to applications in which low supply air temperatures are required. Desiccant cooling is best suited to those applications such as displacement ventilation, where supply air temperatures are close to the room air temperature. Although it is possible to make energy cost savings in 'all air' applications, desiccant cooling systems are best applied to installations in which the bulk of the sensible cooling is performed by a water-based system, such as a chilled ceiling [14].

11.11.1 Solar application of desiccant cooling

Being a heat driven cycle, desiccant cooling affords an opportunity to utilize heat which might otherwise be wasted. It can therefore be coupled to solar collectors to produce a cooling system which, in theory, should be extremely environmentally friendly. However, the use of solar energy puts constraints on the application of desiccant cooling. For example, if the ratio of solar collectors to building floor area is 1:10, then the available heat (in a northern European application) to power the cycle will be in the region of 25 W/m^2 to 50 W/m^2, depending on the climate, type and orientation of the solar collectors [22]. Therefore, if this 'solar' heat is to be harnessed effectively, the desiccant cooling system must be applied in the correct fashion. The desiccant cooling cycle is an open cycle and as such it rejects moist air at a high temperature, which is unsuitable for recirculation. In fact, the greater the air volume flow rate supplied to the room space, the greater the fan power required and the heat energy consumed. Therefore, if desiccant cooling is used in an *all air* application, the regeneration heat load is going to be very large, many times greater than the available solar energy. The bulk of the room sensible cooling should therefore be carried out using a water-based system such as a chilled ceiling, with the desiccant AHU dehumidifying and 'tempering' the incoming fresh air. This strategy reduces the size of the AHU and its associated ductwork, and enables the solar energy to make a significant contribution [22].

References

1. Grigg, P. F. and John, R. W. (1991). Building services technologies to reduce greenhouse emissions. *Proceeding of CIBSE National Conference,* Canterbury, April, p. 231.
2. Calder, K. and Grigg, P. F. (1987). CO_2 impact of refrigeration used for air conditioning. BRE internal report.
3. Wendland, R. D. (1987). Storage to become rule, not exception. *ASHRAE Journal,* May.
4. Energy use in offices (1998). Energy Consumption Guide 19, Department of the Environment, Transport and the Regions.
5. The Accelerated Phase-out of Class I Ozone-Depleting Substances (1999). United States Environmental Protection Agency.

6. Climate Change: Intergovernmental Panel on Climate Change (IPCC) Second Assessment Report (1995).
7. HCFC Phase out Schedule (1998). United States Environmental Protection Agency.
8. Calm, J. M. Global warming impacts of chillers (1993). *IEA Heat Pump Centre Newsletter*, **11**(3), 19–21.
9. Industrial Refrigeration Plant: Energy Efficient Operation and Maintenance (1992). Good Practice Guide 42, Department of the Environment.
10. The Economic Use of Refrigeration Plant, Fuel Efficiency Booklet 11 (1994). Department of the Environment.
11. Jones, W. P. *Air conditioning engineering*, (Chapter 9). Arnold.
12. Energy efficient selection and operation of refrigeration compressors (1994). Good Practice Guide 59, Department of the Environment.
13. Dossat, R. J. (1981). *Principles of refrigeration*. Wiley, p. 436.
14. Beggs, C. B. and Warwicker, B. (1998). Desiccant cooling: Parametric energy study. *Building Services Engineering Research and Technology*, **19**(2), 87–91.
15. Bunn, R. (1993). Learning curve. *Building Services Journal*. October, pp. 20–25.
16. Standeven, M., Cohen, R., Bordass, B. and Leaman, A. PROBE 14: Elizabeth Fry Building. *Building Services Journal*. April pp. 37–42.
17. Bunn, R. (1998). Cool desking. *Building Services Journal*. October, pp. 16–20.
18. Beggs, C. B. (1994). Ice thermal storage: impact on UK carbon dioxide emissions. *Building Services Engineering Research and Technology*, **15**(1), 11–17.
19. Beggs, C. B. and Ward, I. (1992). Ice Storage: design study of the factors affecting installations. *Building Services Engineering Research and Technology*, **13**(2), 49–59.
20. Evaporative cooling applications (1990). *ASHRAE Applications Handbook* (Chapter 50). ASHRAE.
21. Busweiler, U. (1993). Air conditioning with a combination of radiant cooling, displacement ventilation, and desiccant cooling. *ASHRAE Transactions*, p. 503–10
22. Beggs, C. B. and Halliday, S. (1999). A theoretical evaluation of solar powered desiccant cooling in the United Kingdom. *Building Services Engineering Research and Technology*, **20**(3), 113–117.

Bibliography

Beggs, C. B. and Ward, I. (1992). Ice storage: design study of the factors affecting installations. *Building Services Engineering Research and Technology*, **13**(2), 49–59.

Beggs, C. B. (1994). Ice thermal storage: impact on UK carbon dioxide emissions. *Building Services Engineering Research and Technology*, **15**(1), 11–17.

Beggs, C. B. and Warwicker, B. (1998). Desiccant cooling: parametric energy study. *Building Services Engineering Research and Technology*, **19**(2), 87–91.

Beggs. C. B. and Halliday, S. (1999). A theoretical evaluation of solar powered desiccant cooling in the United Kingdom. *Building Services Engineering Research and Technology*, **20**(3), 113–117.

Evaporative cooling applications. (1999). *ASHRAE Applications Handbook* (Chapter 50). ASHRAE.

Jones, W. P. *Air conditioning engineering*. Arnold.

Stoeker, W. F. and Jones, W. (1982). *Refrigeration and air conditioning*. McGraw Hill.

12

Energy efficient electrical services

Much energy is needlessly wasted in buildings through neglect of electrical services. This chapter investigates energy saving measures which can be applied specifically to electrical services in buildings. In particular, low energy lighting and the use of variable speed motor drives are discussed.

12.1 Introduction

Much energy is needlessly wasted in buildings through poor design and maintenance of electrical services. The energy that is wasted is of the worst kind, namely expensive electrical energy, which can be up to five times as expensive as the unit cost of heat. Unfortunately, excessive electrical energy consumption is all too often overlooked by misguided building designers, who focus on thermal energy consumption, which is relatively inexpensive. It has been shown that in a typical 'standard' air conditioned office building in the UK, an average of £3.30 per m^2 (of floor area) per annum is spent running pumps and fans, and a further £2.97 per m^2 is spent on the lighting [1]. This compares with an average of only £1.78 per m^2 spent on heating, and £1.71 per m^2 spent on cooling [1]. These figures demonstrate that in the average office building much more money is spent on running fans, pumps and electric lighting than is spent on operating boilers or refrigeration plant. Yet there are a number of relatively simple technologies that can be applied to motor drives and luminaire installations to dramatically reduce energy costs. That energy costs in these areas can be greatly reduced is clear from the evidence of the UK office building study, which found that in *good practice* standard air conditioned office buildings, only £1.65 per m^2 per year is spent running the pumps and fans, and only £1.48 per m^2 is spent on the lighting [1]. This equates in each case to energy cost reductions of about 50% when compared with typical air conditioned office buildings.

12.2 Power factor

Electric induction motors and fluorescent lamp fittings are classic examples of reactive (i.e. inductive) electrical loads. Reactive electrical loads are important because, unlike resistive loads such as incandescent light, they cause the current to become out of phase with the voltage (see Figure 12.1). This, in simple terms, means that items of equipment which are inductive in nature

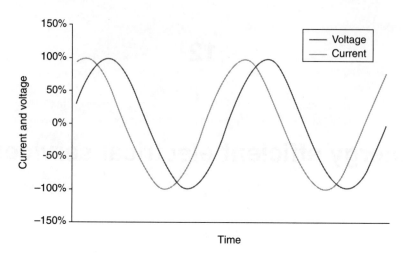

Figure 12.1   The effect of an inductive load on an electrical current and voltage

draw a larger current than would be anticipated by their useful power rating. Ultimately, it is the consumer who has to pay for this additional current.

The electrical power consumed by a resistive load can be determined by:

$$W = V \times I \tag{12.1}$$

where W is the power (W), V is the voltage (V), and I is the current (A).

Equation (12.1) defines the useful power consumed, and applies to all types of resistive load where the current is in phase with the voltage. However, eqn (12.1) does not hold true for reactive loads, where the current lags behind the voltage, since reactive loads consume more power than can be usefully used. A reactive load, such as an induction motor, will therefore draw a larger current than would be anticipated by its useful power rating. The reactive components of the load consume what is termed *reactive power*. In order to determine the apparent power consumed by a reactive load, the true power must be added vectorally to the reactive power, as shown in Figure 12.2.

It should be noted from Figure 12.2 that the reactive power is drawn at right angles to the true power. The apparent power is therefore a function of the true power consumed and the reactive power, and can be expressed as:

$$\text{Apparent power} = \frac{\text{True power}}{\cos\phi} \tag{12.2}$$

where $\cos\phi$ is the power factor.

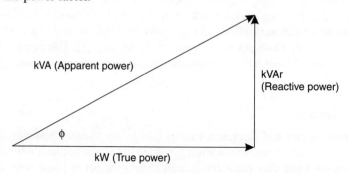

Figure 12.2   The relationship between kW, kVA and kVAr

From Figure 12.2 and eqn (12.2) it is evident that when the current and voltage are in phase with each other (i.e. a resistive load), the apparent power is the same as the true power. When the two are out of phase (i.e. a reactive load) the apparent power consumed is always going to be greater than the true power. In order to differentiate between true and apparent power, true power is measured in watts (W) or kilowatts (kW), and apparent power is measured in volt amps (VA) or kilovolt amps (kVA). Similarly, reactive power is measured in volt amps reactive (VAr) or kilovolt amps reactive (kVAr). The ratio of true power to apparent power is known as the *power factor*. For a pure resistive load the power factor would be 1, and for a pure inductor the power factor would be 0.

$$\text{Power factor} = \frac{\text{True power}}{\text{Apparent power}} \qquad (12.3)$$

12.2.1 Effects of a poor power factor

In many buildings and other installations the overall electrical load is heavily influenced by the presence of reactive loads such as induction motors and fluorescent tubes which create a lagging power factor. As a result power factors of 0.7 or less are often experienced. Example 12.1 illustrates the impact of such a poor power factor.

Example 12.1

A 240 V single-phase electric motor has a true power of 1.8 kW and exhibits a power factor of 0.7. Determine:

(i) The current required to drive the motor.
(ii) The current required if the power factor was 1.

Solution

$$\text{True power (W)} = \text{Apparent power (VA)} \times \text{Power factor}$$

Therefore

$$\text{Current} = \frac{\text{Watts}}{\text{Volts} \times \text{Power factor}}$$

(i)

$$\text{Current} = \frac{1800}{240 \times 0.7} = 10.71 \text{ A}$$

(ii)

$$\text{Current} = \frac{1800}{240 \times 1} = 7.5 \text{ A}$$

Example 12.1 demonstrates that the lower the power factor, the greater the current required to provide the same useful power. The increased current required as a result of a poor power factor has the knock-on effect of increasing power losses. Because cables and other items of equipment have an electrical resistance, power is lost as heat when a current flows. The power (or I^2R) loss can be expressed as:

$$\text{Power loss} = I^2 \times R \qquad (12.4)$$

where I is the current (A), and R is the resistance (Ω).

It can be seen that for a circuit with a constant resistance, the greater the current, the greater the I^2R losses. In addition to increased currents and increased I^2R losses, a poor power factor has the knock-on effect that switchgear, cables and transformers all have to be increased in size. Example 12.2 illustrates this fact.

Example 12.2

A building is served by a 415 V (line-to-line) three phase supply and has a true power load of 210 kW and a power factor of 0.7. Compare the installation with a similar one having a power factor of 1.

Solution

$$\text{Total current} = \frac{210\ 000}{415 \times 0.7 \times \sqrt{3}} = 417.4\ \text{A}$$

If however, the power factor was 1, then

$$\text{Total current} = \frac{210\ 000}{415 \times 1 \times \sqrt{3}} = 292.2\ \text{A}$$

The impact of this reduction in current is shown in Table 12.1.

Examples 12.1 and 12.2 show that the lower the power factor the greater the current drawn and the greater the size of the infrastructure required. Therefore, if a consumer has a poor power factor the electricity utility company has to supply more 'electricity' than will be recorded as 'true power' in kW at the electricity meter. This means that the utility company will not be paid in full for all the electricity which it is supplying to the consumer. Utility companies overcome this problem by adopting one of two strategies:

Strategy 1: Installing meters, and offering tariffs which record electricity consumption in kVA and not in kW.

Strategy 2: Using meters and tariffs which record electricity consumption in kW and levy an additional charge for the number of reactive power units (kVAr) consumed.

By using either approach a utility company can ensure that it receives the correct revenue for the electrical energy it supplies.

12.2.2 Power factor correction

The simplest way to correct a poor power factor is to minimize the problem in the first place. In many applications a poor power factor occurs as a result of the use of induction motors. Induction motors are commonplace in buildings and are used to drive fans and pumps. As such they are a necessity and cannot be avoided. The power factor of induction motors varies with the

Table 12.1 Impact of poor power factor

	Power factor = 0.7	*Power factor* = 1
Total current	417.4 A	292.2 A
Apparent power (kVA)	300 kVA	210 kVA
Switchgear rating	450 A	350 A
Transformer rating	400 kVA	300 kVA
Cable size	240 mm²	150 mm²

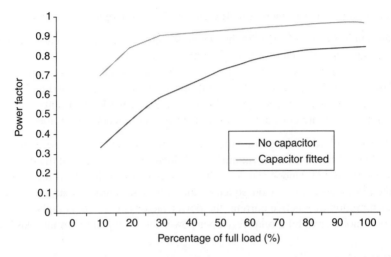

Figure 12.3 Impact of power factor correction on induction motors[2]

motor loading. Motors which may have a power factor of 0.8 at full load may have a power factor approaching 0.1 at low load, with the result that almost 90% of the total current drawn is reactive in nature [2]. Motors should therefore be selected with care, since an under loaded large motor will exhibit a low power factor. The power factors exhibited by smaller motors are not as good as those of larger motors. Despite this, it is usually better to select a smaller motor than use an under-loaded large motor to perform the same job.

It is possible to correct a poor power factor by installing capacitors. The effect of capacitors on an alternating current is the opposite to that of a reactive load. They cause the current to lead the voltage. By installing capacitors into an electrical circuit it is possible to counteract the effect of any reactive load and correct a poor power factor. Power factor correcting capacitors can either be installed in a central bank before the main distribution panel, or mounted on individual items of equipment. It is generally considered better to correct the power factor for reactive loads at the item of equipment itself. This reduces the current drawn by the item of equipment and thus reduces the I^2R losses in all the wiring leading to the item of equipment. By fitting capacitors to an induction motor it is possible to greatly improve its power factor. An example of this is shown in Figure 12.3, where the introduction of capacitors results in the power factor being virtually constant with all loads over 50% of full load [2].

In large installations it may be more cost effective to install a central bank of capacitors to correct power factor. It is possible to use banks of capacitors which automatically switch on and off in order to maintain an optimum power factor.

12.3 Electric motors

Induction motors are widely used in many applications. Pumps, fans, compressors, escalators and lifts are all powered by motors of one type or another. Induction motors are therefore essential to the operation of most modern buildings. Furthermore, electric motors are often the most costly items of plant to run in many office buildings. It is therefore well worth understanding how induction motors use electrical energy and investigating possible energy conservation measures.

All induction motors have inherent inefficiencies. These energy losses include [3]:

- Iron losses which are associated with the magnetic field created by the motor. They are voltage related and therefore constant for any given motor and independent of load.
- Copper losses (or I^2R losses) which are created by the resistance of the copper wires in the motor. The greater the resistance of the coil, the more heat is generated and the greater the power loss. These losses are proportional to the square of the load current.
- Friction losses which are constant for a given speed and independent of load.

These losses can be divided into those which vary with motor load and those which are constant whatever the load. When a motor is running at full load, the split between the two is about 70% and 30% respectively [3]. Under part load this split changes; at low load the current drawn is small and the I^2R losses are low. Consequently, the iron losses predominate and since they result from the consumption of reactive current, the power factor is correspondingly low. Even at full load, induction motors exhibit a relatively poor power factor, typically around 0.8 [3].

12.3.1 Motor sizing

Correct sizing of electric motors is critical to their efficient operation, since oversized motors tend to exhibit poor power factors and lower efficiencies. Depending on size and speed, a typical standard motor may have a full load efficiency between 55% and 95% [2]. Generally, the lower the speed, the lower the efficiency, and the lower the power factor. Typically motors exhibit efficiencies which are reasonably constant down to approximately 75% full load. Thereafter they may lose approximately 5% down to 50% of full load, after which the efficiency rapidly falls (as shown in Figure 12.4) [2].

It can be seen from the performance curve in Figure 12.4 that it is possible to oversize a motor by up to 25% without seriously affecting its efficiency, provided that a motor is run at a relatively constant load. If the load fluctuates and rarely achieves 75% full load, then both the efficiency and the power factor of the motor will be adversely affected. In fact the power factor tends to fall off more rapidly than the efficiency under part-load conditions. Therefore, if motors are oversized, the need for power factor correction becomes greater. Oversizing of motors also increases the capital cost of the switchgear and wiring which serves the motor.

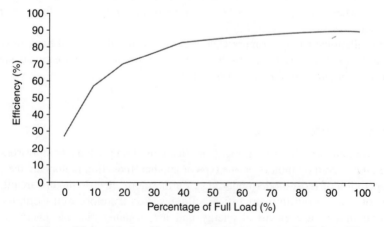

Figure 12.4 Relationship between motor loading and efficiency[2]

12.4 Variable speed drives (VSD)

Most induction motors used in buildings are fitted to fans or pumps. The traditional approach to pipework and ductwork systems has been to oversize pumps and fans at the design stage, and then to use commissioning valves and dampers to control the flow rate by increasing the system resistance. While mechanical constrictions are able to control the flow rate delivered by fans and pumps (see Figure 12.5), the constriction itself increases the system resistance and results in increased energy loss. This situation is highly undesirable and is one of the main reasons why the energy consumption associated with fans and pumps is so high in so many buildings [1]. An alternative approach to the use of valves and dampers is to control the flow rate by reducing the speed of the fan or pump motor. This strategy results in considerable energy savings, as illustrated in Example 12.3.

Example 12.3

It is proposed to use a forward-curved centrifugal fan in a mechanical ventilation system. The fan is required to deliver a volume flow rate of 1.8 m³/s and the estimated system resistance is 500 Pa. However, the proposed fan delivers 2.06 m³/s against a resistance of 500 Pa while running at a speed of 1440 rpm. Determine the fan power input, if:

(a) A volume control damper is used to achieve a volume flow rate of 1.8 m³/s by increasing the total system resistance to 750 Pa.
(b) The fan speed is reduced in order to deliver 1.8 m³/s.

Solution

(a) Fan air power input:

$$W = \dot{v} \times P_{\mathrm{t}}$$

where $\dot{v}$ is the air volume flow rate (m³/s), and P_{t} is the total system resistance (Pa).
Let W_1 be the fan power when delivering 2.06 m³/s against a resistance of 500 Pa, and W_2 be the fan power when delivering 1.8 m³/s against a resistance of 750 Pa.

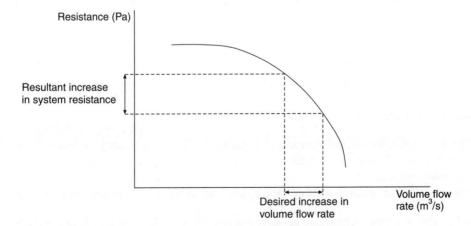

Figure 12.5   Impact of a volume control damper on system resistance

Therefore

$$W_1 = 2.06 \times 500 = 1030 \text{ W}$$

and

$$W_2 = 1.8 \times 750 = 1350 \text{ W}$$

Therefore

$$\text{Increase in power consumption} = \frac{1350 - 1030}{1030} \times 100 = 31.1\%$$

(b) The fan laws state that:

$$\dot{v} \, \alpha \, N$$

and

$$W \, \alpha \, N^3$$

where v is the air volume flow rate (m³/s), N is the fan speed (rpm), and W is the fan air power input (W).

Let N_1 be the fan speed when delivering 2.06 m³/s against a resistance of 500 Pa, N_3 be the fan speed when delivering 1.8 m³/s, and W_3 be the fan power when delivering 1.8 m³/s.

Therefore

$$N_3 = 1440 \times \frac{1.8}{2.06} = 1258.3 \text{ rpm,}$$

$$W_3 = 1030 \times \frac{1.8^3}{2.06^3} = 687.2 \text{ W.}$$

Therefore

Reduction in power consumption (W^3 compared with W^1) $= \dfrac{1030 - 687.2}{1030} \times 100 = 33.3\%$

However

Reduction in power consumption (W^3 compared with W^2) $= \dfrac{1350 - 687.2}{1350} \times 100 = 49.1\%$

It can be seen from Example 12.3 that:

- The use of volume control dampers to regulate air flow significantly increases fan energy consumption. The precise magnitude of this increase will depend on the characteristics of the particular fan selected.
- Reducing the fan speed to regulate the air flow rate always results in fan energy savings.

The fan power savings which can be achieved through reducing fan speeds are considerable, especially when compared with the fan power increase which results from using volume control dampers. As a result there are great advantages to be gained, if fan and pump speeds can be controlled.

The energy savings achieved in Example 12.3 are indicative of the type of savings which can be achieved through the use of VSDs on fans and pumps. In most applications the potential for saving energy through the use of VSDs on pumps, fans and compressors is considerable. Most designers overestimate system resistances with the result that most pumps and fans are theoretically oversized before the actual fan or pump selection is undertaken. During the selection process, the cautious designer is unlikely to find a fan, or pump, which matches the theoretical 'calculated' specification and thus a larger one is selected which is sure to perform the required task. This strategy protects the system designer and ensures that he/she does not negligently undersize the fans or pumps. Unfortunately, it also ensures that the system is greatly oversized and that during the commissioning process, volume control dampers and dampers will have to be used to reduce the volume flow rate. Consequently, both the capital and future operating costs of the system are greatly increased. By using VSDs it is possible to ensure that even if fans and pumps are oversized, energy consumption will not be greatly increased. This makes the installation of VSDs one of the most cost effective energy efficiency measures that can be taken. It has been estimated that for VSDs payback periods of less than 2 years are the norm [2].

In addition to the energy savings gained through using VSDs on constant flow systems, even greater savings can be made by employing VSDs on variable volume flow systems. When the load profiles and duty cycles of heating, air conditioning and ventilation systems are examined in detail, it is found that most regularly operate well below their intended design specification. The main reason for this is that system designers are overcautious at the design stage. As a result, over-large constant volume flow rate, variable temperature systems are designed. While this approach works in practice, it means that pump and fan running costs are constant and high, no matter what the operating load. An alternative approach is to keep the temperature constant and vary the flow rate, so that pump and fan running costs reduce as the operating load reduces. The classic system which adopts this approach is the VAV air conditioning system, for which VSDs are ideally suited.

12.4.1 Principles of VSD operation

Modern electronic VSD systems adjust the mains alternating current to regulate motor speed. Various electronic VSD systems are available. One of the most popular types is the *variable frequency drive*, which achieves speed control by varying the voltage and frequency output. Such drives regulate the voltage to the motor in proportion to the output frequency in order to ensure that the ratio of voltage to frequency remains relatively constant. Changes in motor speed are achieved by modulating the voltage and frequency to the motor. Figure 12.6 shows the basic components in a *variable frequency drive* VSD system.

Variable frequency drive systems comprise two main components, a rectifier and an inverter. The rectifier converts standard alternating current (ac) (e.g. 240 V and 50 Hz) to an adjustable direct current (dc), which is then fed to the inverter. The inverter comprises electronic switches which turn the dc power on and off to produce a pulsed ac power output. This can then be controlled to produce the required frequency and voltage. The switching characteristics of the inverter are modified by a regulator, so that the output frequency can be controlled.

The inverter is the critical part of a VSD system. One type of inverter currently in use is the pulse width modulated (PWM) inverter, which receives a fixed dc voltage from the rectifier and adjusts the output voltage and frequency. The PWM inverter produces a current waveform which approximates to the pure sine wave of mains ac supply.

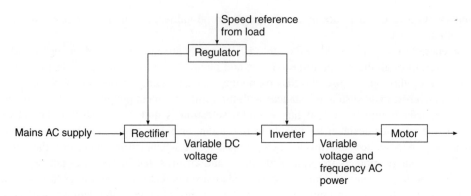

Figure 12.6   Components of a variable speed drive. Crown copyright is reproduced with the permission of the Controller of Her Majesty's Stationery Office and the Queen's Printer for Scotland [4]

12.5 Lighting energy consumption

The energy consumed by electric lighting in most building types is considerable. Table 12.2 shows the proportion of overall energy consumed by lighting for a variety of building types in the UK.

Although in many buildings the energy consumed by the heating system is often greater than that consumed by lighting, the energy costs associated with lighting are often considerably greater than those associated with the heating [1]. It is possible to achieve considerable energy cost savings through the careful design and maintenance of lighting schemes. On average, *good practice* 'standard' air conditioned office buildings in the UK experience an annual lighting cost is £1.48 per m², which compares very favourably with the typical value of £2.97 per m² of floor space [1].

12.5.1 Daylighting

Although the focus of this chapter is on electrical services, daylighting is relevant to the subject of artificial lighting and so a short discussion is included here. The ability of daylighting to reduce lighting energy costs should not be underestimated. Daylight can make a substantial contribution to the lighting of buildings by reducing reliance on artificial lighting.

The major factors affecting the daylighting of an interior are the depth of the room, the size and location of windows, the glazing system, and any external obstructions. These factors usually depend on decisions made at the initial design stage. Through appropriate planning at an early stage it is possible to produce a building which is energy efficient as well as having a pleasing internal appearance. Glazing can, however, impose severe constraints on the form and

Table 12.2   Typical energy consumption on lighting for various applications in the UK [5]

Building type	Typical percentage of energy consumed by lighting (%)
Banks	19
Factories	15
Hotels	9
Offices	16–20
Schools	9–12
Supermarkets	11

operation of a building. If poor design decisions are made concerning fenestration it is possible to create a building in which the occupants are uncomfortable, and in which energy consumption is high. Glazing should therefore be treated with care!

12.5.2 Lighting definitions

Before discussing the factors which influence the energy consumption of artificial lighting schemes, it is important first to understand the terminology involved, and to appreciate how lighting schemes are designed. A full discussion of the subject of lighting design is however beyond the scope of this book.

When an incandescent lamp is switched on, it emits a luminous flux in all directions. The fundamental SI unit of *luminous flux* is the lumen (lm). It should be noted that the lumen is simply a measure of the quantity of luminous flux. It tells us nothing about the direction of the light. When a lamp is placed in a luminaire fitting with an integral reflector, the luminous flux from the lamp will be directed in one particular direction (e.g. downward in the case of a ceiling mounted fitting). A certain number of lumens are therefore focused in a particular direction with a certain *luminous intensity*. Luminous intensity, as the term suggests, is the intensity of luminous flux in any given 3-dimentional angular direction and its SI unit is the candela (cd). '3-dimensional angular direction' is a difficult concept to define; but it is usually referred to as *solid angle*. It is the 3-dimensional equivalent of a 2-dimensional angle. The steradian is the SI unit of *solid angle* and is the 3-dimensional equivalent of the radian. The candela can therefore be defined as being a lumen per steradian. More precisely, one candela can be defined as the luminous intensity from a source producing light at 540 000 000 MHz and at a specific intensity of 1/683 W per steradian [6].

Lighting manufacturers use a system of polar diagrams to describe luminous intensity distribution from luminaire fittings. Figure 12.7 shows a typical polar diagram for a transverse section across a ceiling mounted luminaire fitting. It should be noted that luminous intensity produced by the lamp is not the same in all directions. For example, while the intensity in a vertical direction is 112 cd per klm of lamp flux, the intensity at 40° to the vertical is only 88 cd per klm. Lighting manufacturers usually specify intensity in terms of cd per klm of installed lamp flux because it is often possible to use a variety of lamps in any particular luminaire fitting.

From a purely functional point of view, it is not so much the intensity of light coming from a light source (i.e. a luminaire fitting) that is important, but rather the amount of light that is falling on a particular surface. For example, a room may be illuminated only by spotlights which shine brightly on certain specific objects while leaving the rest of the room in relative darkness. If a person tries to read a book in a region of the room which is not well lit, they will experience difficulties. Although there may be a large luminous flux in the room, the problem is that very little of it is falling on the pages of the book. The person then experiences difficulties in reading. The amount of luminous flux falling on a surface is therefore of great importance. It is referred to as *illuminance* and has the SI unit, the lux (lx). One lux is defined as a luminous flux of 1 lumen falling on a surface having an area of 1 m². Artificial lighting schemes are usually specified as being capable of supplying a specified number of lux on a horizontal working plane. Generally, the more demanding the work, the higher the level of illuminance required.

Because light is a form of radiant energy, its ability to illuminate a surface (i.e. its illuminance) varies inversely with the square of the distance between the source and the surface. In simple terms, if the distance increases, the illuminance decreases by the square of the distance. This

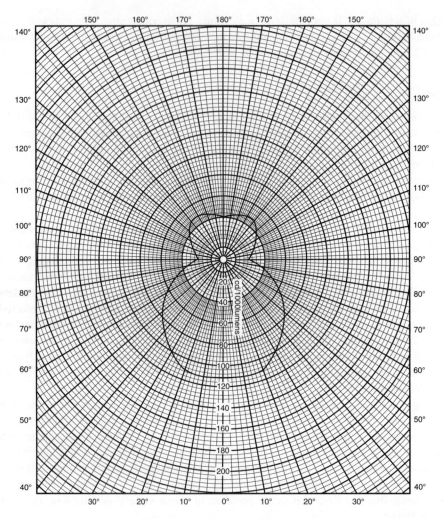

Figure 12.7　Polar intensity diagram (from Smith, Phillips and Sweeney (1987) *Environmental Science* © Longman Group Ltd, reprinted by permission of Pearson Education Ltd.)

relationship is known as the inverse square law. Also, if a horizontal surface is illuminated from the side, so that the light hits the surface at an angle other than 90°, the available luminous flux is shared out over a larger surface area so that illuminance decreases. The angle at which light strikes a surface is therefore of importance. This fact can be combined with the inverse square law to produce the cosine law of illuminance, which can be expressed as

$$\text{Illuminance on a horizontal surface } E_h = \frac{I_\theta}{d^2} \times \cos\theta \qquad (12.5)$$

where θ is the angle at which light strikes the horizontal surface (i.e. angle from the vertical) (°), I_θ is the luminous intensity in direction θ (cd), and d is the distance of plane from light source (m).

Example 12.4 shows how the cosine law can be applied to the data contained in the polar diagram shown in Figure 12.7.

Example 12.4

Using the luminaire polar diagram shown in Figure 12.7, determine the illuminance on a horizontal surface: (Assume that the luminaire fitting contains 2 fluorescent tubes each with a luminous flux of 3200 lm.)

(i) At a point (a) 3 m directly below the luminaire.
(ii) At a point (b) 2 m to the right of point (a).

Solution

(i) From Figure 12.7 it can be seen that the luminous intensity in the vertical direction is 112 cd per klm of lamp flux. Therefore

$$\text{Luminous intensity in the vertical direction} = 112 \times (2 \times 3.2)$$
$$= 716.8 \text{ cd.}$$

Therefore:

$$\text{Illuminance on a horizontal surface at point (a)} = \frac{716.8}{3^2} \times \cos\theta = 79.6 \text{ lx}$$

(ii) Considering point (b)

$$\tan\theta = \frac{2}{3} = 0.667$$

Therefore

$$\theta = 33.69°$$

and

$$d = \overline{(2^2 + 3^2)} = 3.606 \text{ m}$$

From Figure 12.7 it can be seen that the luminous intensity at 33.7° from in the vertical is 96 cd per klm of lamp flux. Therefore

$$\text{Luminous intensity in the vertical direction} = 96 \times (2 \times 3.2)$$
$$= 614.4 \text{ cd}$$

Therefore

$$\text{Illuminance on a horizontal surface at point (b)} = \frac{614.4}{3.606^2} \times \cos 33.69 = 39.3 \text{ lx}$$

12.6 Artificial lighting design

The performance of an artificial lighting scheme is influenced by:

- The efficacy of the lamps (i.e. the light output per watt of electrical power consumed).
- The luminaire performance.

- The layout of the luminaire fittings.
- The surface reflectance of the decor and furnishing.
- The maintenance standards.

All these factors have to be allowed for when designing any lighting scheme. One method which is frequently used and which considers all these factors is the *lumen design method*. The lumen method enables regular lighting schemes to be designed quickly and easily, and so is particularly popular as a design method. The method enables the number of luminaires to be determined for any rectangular room space using eqn (12.6).

$$\text{Number of luminaire fitting required } n = \frac{E_{av} \times A}{\phi \times UF \times MF} \tag{12.6}$$

where n is the number of luminaire fittings required, A is the area of working plane in room (i.e. room area) (m^2), E_{av} is the average illuminance required on the working plane (lx), ϕ is the lighting design lumens per fitting (lm), UF is the utilization factor of luminaire fitting, and MF is the maintenance factor.

Each of the terms in eqn (12.6) corresponds with list of factors outlined at the beginning of this section. However, in order to understand the relevance of each term, some explanation is required.

12.6.1 Average illuminance (E_{av})

The required illuminance in a room depends on the nature of the tasks being undertaken in the space. Visual acuity improves at higher levels of illuminance. The more visually demanding the task, the higher the level of illuminance required on the working plane. The working plane is normally taken to be desk height. Table 12.3 shows appropriate levels of illuminance for a variety of activities and spaces.

12.6.2 Lighting design lumens (ϕ)

The term *lighting design lumens* simply refers to the total lumen output of the lamps in a particular luminaire fitting. It should be noted that this is the lumen output when the lamps are new. It

Table 12.3   Required standard illuminances for various activities [7]

Standard maintained illuminance (lx)	Representative activities
50	Cable tunnels, indoor storage tanks, walkways
100	Corridors, changing rooms, bulk stores, auditoria
150	Loading bays, medical stores, switch rooms, plant rooms
200	Entrance foyers, monitoring automatic processes, casting concrete, turbine halls, dining rooms
300	Libraries, sports and assembly halls, teaching spaces, lecture theatres
500	General office spaces, engine assembly, kitchens, laboratories, retail shops
750	Drawing offices, meat inspection, chain stores
1000	General inspection, electronic assembly, gauge and tool rooms, supermarkets
1500	Fine work and inspection, hand tailoring, precision assembly
2000	Assembly of minute mechanisms, finished fabric inspection

is also important to appreciate that lighting manufacturers produce luminaire fittings which might accommodate a variety of lamps, each of which will emit a different luminous flux. The type of lamp to be used in the luminaire should therefore be specified.

12.6.3 Utilization factor (UF)

The UF can be expressed as:

$$UF = \frac{\text{Total flux reaching the working plane}}{\text{Total lamp flux}} \qquad (12.7)$$

The UF takes into account both the direct luminous flux which reaches the working plane straight from the luminaire and the flux which reaches the working plane having been reflected from the walls and the ceiling. The UF is influenced by the nature of the luminaire used, the room surface reflectance, and the room dimensions. Table 12.4 shows a typical set of UF data produced by a manufacturer for a particular luminaire fitting.

It should be noted that both the room surface reflectance and the room geometry are allowed for in Table 12.4. The room geometry is allowed for by the 'room index' which is expressed as:

$$\text{Room index} = \frac{L \times W}{H_m \times (L + W)} \qquad (12.8)$$

where L is the length of room (m), W is the width of room (m), and H_m is the mounting height (i.e. height above working plane) (m).

12.6.4 Maintenance factor (MF)

The effective light output from any luminaire decreases with time. This is because of a number of factors:

- Lamp output decreases with time.
- Luminaire reflectors and diffusers become dirty with time.
- Room surfaces become dirty with time.

To compensate for this drop in luminaire output, a MF is introduced into the lumen design method. The MF also allows for the fact that periodically individual lamps fail and remain unattended for some period of time until they are replaced.

Both lamp survival and lamp output fall with time. Figure 12.8 shows output and survival characteristic curves for fluorescent and tungsten filament lamps. It can be seen that two types of lamps behave very differently. The lumen output of a fluorescent lamp falls by nearly 10% during the first 500 hours of operation [7]. Thereafter, the output decreases less rapidly. With a tungsten filament lamp the decrease in light output is much more gradual, although the lamp life is much shorter than that of a fluorescent tube.

Table 12.4   Luminaire utilization factor table

Room reflectance's ceiling (%) [wall %]		1.00	1.25	1.50	2.00	2.50	3.00
				Room index			
Room reflectance's	50 [50]	0.50	0.54	0.57	0.61	0.63	0.65
ceiling (%) [wall %]	50 [30]	0.47	0.51	0.54	0.58	0.61	0.63
	50 [10]	0.44	0.48	0.51	0.56	0.59	0.61

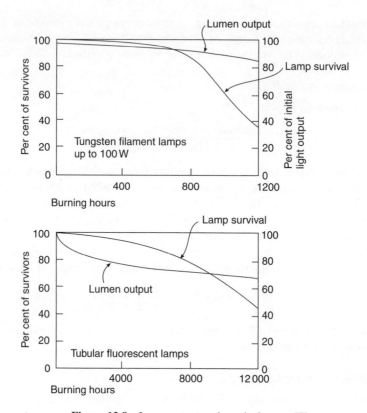

Figure 12.8   Lamp output and survival curves[7]

The extent to which luminaire fittings become dirty depends very much on the type of fitting. For example, a ventilated luminaire, used in conjunction with an extract plenum, will cause dust and dirt from the room space to collect on the luminaire reflector with the result that it may quickly become dirty. Figure 12.9 shows how light output decreases with dirt deposition for a variety of luminaire types.

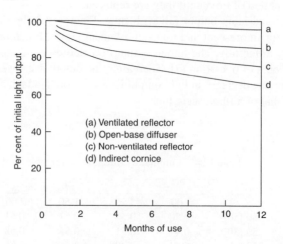

Figure 12.9   Fall in light output due to dirt deposition [7]

Tables 12.5, 12.6 and 12.7 are attempts to quantify the various factors associated with decreased luminaire output. Table 12.5 quantifies lamp performance over time in two ways, the *lumen maintenance factor* (LLMF) is the proportion of initial lamp lumens 'remaining' after a given time period, while the *lamp survival factor* (LSF) is the proportion of lamps surviving (i.e. lamps that have not failed) after a given time [8].

The *luminaire maintenance factor* (LMF) referred to in Table 12.6 quantifies the impact of various maintenance regimes on different luminaire types in a variety of environments.

The impact of various cleaning regimes on room surface reflectance is quantified in Table 12.7 by using a *room surface maintenance factor* (RSMF).

Example 12.5 shows how a realistic MF might be developed in practice.

Example 12.5

The lighting scheme in an office space comprises 60 batten type luminaire fittings with fluorescent (tri-phosphor type) tubes. The lamps are changed in bulk every 6000 hours and the luminaire fittings and room surfaces are cleaned every 12 months. The cleanliness of the environment within the office space is normal and the luminous flux distribution

Table 12.5   Lamp lumen maintenance and survival factors

Lamp type	Factor	Operating hours		
		6000	10 000	12 000
Fluorescent (tri-phosphor)	LLMF	0.87	0.85	0.84
	LSF	0.99	0.85	0.75
Metal halide	LLMF	0.72	0.66	0.63
	LSF	0.91	0.83	0.77
Sodium (high-pressure)	LLMF	0.91	0.88	0.87
	LSF	0.96	0.92	0.89

From Pritchard (1995) *Lighting* © Longman Group UK Ltd, reprinted by permission of Pearson Education Ltd

Table 12.6   Luminaire maintenance factor (LMF)

Luminaire type	6 months cleaning interval			12 months cleaning interval			18 months cleaning interval		
	Clean	Normal	Dirty	Clean	Normal	Dirty	Clean	Normal	Dirty
Batten	0.95	0.92	0.88	0.93	0.98	0.83	0.91	0.87	0.80
Enclosed IP2X	0.92	0.87	0.83	0.88	0.82	0.77	0.85	0.79	*
Up-lighter	0.92	0.89	0.85	0.86	0.81	*	0.81	*	*

* Not recommended.
From Pritchard (1995) *Lighting* © Longman Group UK Ltd, reprinted by permission of Pearson Education Ltd

Table 12.7   Room surface maintenance factor (RSMF)

Room index	Luminaire flux distribution	12 months cleaning interval			24 months cleaning interval		
		Clean	Normal	Dirty	Clean	Normal	Dirty
2.5–5.0	Direct	0.98	0.96	0.95	0.96	0.95	0.94
2.5–5.0	General	0.92	0.88	0.85	0.89	0.85	0.81
2.5–5.0	Indirect	0.88	0.82	0.77	0.84	0.77	0.70

From Pritchard (1995) *Lighting* © Longman Group UK Ltd, reprinted by permission of Pearson Education Ltd

from the luminaries is 'general' in nature. Determine a suitable maintenance factor for the installation.

Solution

From Table 12.5

$$LLMF = 0.87$$
$$LSF = 0.99$$

From Table 12.6

$$LMF = 0.98$$

From Table 12.7

$$RSMF = 0.88$$

Therefore

$$MF = 0.87 \times 0.99 \times 0.98 \times 0.88 = 0.74$$

When designing a lighting scheme it is important to ensure that the illuminance on the working plane is evenly distributed. If the luminaires are too far apart, gloomy patches will appear on the working plane. This makes it important not to exceed the *spacing to mounting height ratio* stated for the particular luminaire being used. The nominal spacing for luminaire fittings can be determined using eqn (12.9).

$$\text{Nominal spacing between fittings } S = \sqrt{\frac{A}{n}} \qquad (12.9)$$

The process involved in the lumen design method is illustrated in Example 12.6.

Example 12.6

A 15 × 9 × 3 m high office space is required to be illuminated to 500 lux. Given the following data design a suitable artificial lighting layout for the space.

Data:

Height of working plane = 800 mm
Height of luminaire fittings above floor level = 3000 mm
Design lumens per fitting = 8000 lm
Maintenance factor = 75%
Ceiling reflectance factor = 50%
Wall reflectance factor = 50%
Maximum spacing to mounting height ratio = 1.2 : 1.0

		Room index					
		1.00	1.25	1.50	2.00	2.50	3.00
Room reflectance's	50 [50]	0.50	0.54	0.57	0.61	0.63	0.65
ceiling (%) [wall %]	50 [30]	0.47	0.51	0.54	0.58	0.61	0.63
	50 [10]	0.44	0.48	0.51	0.56	0.59	0.61

Solution

Mounting height $= 3.0 - 0.8 = 2.2$ m

Using eqn (12.6)

$$\text{Room index} = \frac{15 \times 9}{2.2 \times (15 + 9)} = 2.56$$

Since, the ceiling reflectance is 50%, the wall reflectance is 50% and the room index is 2.56. Therefore

$$UF = 0.632$$

Using eqn (12.6)

$$\text{Number of luminaire fitting required } n = \frac{500 \times (15 \times 9)}{8000 \times 0.632 \times 0.75} = 17.8$$

Since it is impossible to have 0.8 of a luminaire fitting, the number of luminaires required (n) must be 18.

Using eqn (12.9)

$$\text{Nominal spacing between fittings } S = \sqrt{(15 \times 9)/18} = 2.739 \text{ m}$$

Therefore

$$\text{Spacing:Mounting height} = 2.739{:}2.2 = 1.245{:}1.0$$

However, this spacing to mounting height ratio exceeds the maximum permissible ratio of 1.2:1.0. Therefore, it is necessary to increase the number of fittings to, say 20.

Therefore

$$\text{Nominal spacing between fittings } S = \sqrt{(15 \times 9)/20} = 2.598 \text{ m}$$

Therefore

$$\text{Spacing:Mounting height} = 2.598{:}2.2 = 1.181{:}1.0$$

This spacing to mounting height ratio is acceptable. Therefore

Number of luminaire fittings required $= 20$

and

Suggested design layout: 4 rows of 5 fittings

12.7 Energy efficient lighting

The main factors which influence the energy consumption of lighting schemes are:

(i) The light output per watt of electrical power consumed (i.e. lamp efficacy).
(ii) Luminaire performance.
(iii) The number of luminaires and their location.
(iv) The reflectance of internal room surfaces.
(v) Maintenance and procedure standards.
(vi) Duration of operation.
(vii) The switching and control techniques used.

While points (i) to (iv) above are concerned with the fundamental efficiency of any installation, points (v) to (vii) relate to the management and operation of the installation.

When considering the overall energy efficiency of luminaires it is helpful to look in isolation at the individual components which together make-up the luminaire; namely the lamp, the control gear and the fitting.

12.7.1 Lamps

There are a wide variety of lamps which can be used in artificial lighting schemes. Fluorescent, tungsten filament, tungsten halogen, metal halide (MBI) and high pressure sodium vapour (SON) are amongst the many lamp types in common use. Energy consumption varies greatly with the type of lamp used. Figure 12.10 shows comparative lamp efficacies for a variety of lamp types. Luminous efficacy is defined in lumens produced per watt of electricity consumed.

It can clearly be seen from Figure 12.10 that there is a wide variation in the luminous efficacy between the various lamp types. For example, compact fluorescent lamps have an efficacy of approximately 70 lm/W, while tungsten filament lamps exhibit an efficacy of approximately 10 lm/W. Clearly, the compact fluorescent lamp is a much more energy efficient option. It should also be noted that the luminous efficacy of any lamp type increases as power input is increased. While this increase may only be slight in some types of lamp (e.g. tungsten filament and tungsten halogen lamps), in others, such as high and low pressure sodium lamps (SON and SOX), the increase can be substantial.

When designing new artificial lighting installations, it is important to install lamps which exhibit a high efficacy. In older installations, it may be worth considering refurbishing existing luminaire fittings, so that they can incorporate newer more efficient lamp types. Refurbishment of older installations using modern equipment can often result in substantial energy savings as well as improved visual conditions. It is possible to improve older, less efficient, luminaires by replacing existing diffusers with modern reflector systems at relatively low cost. However, some changes can be considerably more expensive, such as replacement of the existing control gear to facilitate the use of low energy lamps.

Tungsten filament lamps

Although the use of tungsten filament lamps is widespread, they are particularly inefficient consumers of energy and should be avoided where possible. They have efficacies in the region of 8 to 15 lm/W [9], with most of the electrical energy being converted to heat, which can lead to space overheating problems. Lamp life is short, with most tungsten filaments burning out after approximately 1000 hours of use [7].

Compact fluorescent lamps

Because standard tungsten filament lamps exhibit such poor efficacies, compact fluorescent lamps were developed as a replacement. Where possible tungsten lamps should be replaced by compact fluorescent lamps. These give comparable light output to tungsten lamps, but only consume approximately 20% of the power required by tungsten lamps [8]. As the rated life of the compact fluorescent lamps is in the region of 8000–12 000 hours, eight times longer than tungsten lamps, maintenance costs are greatly reduced, albeit at a higher initial cost.

Compact fluorescent lamps can be divided into two distinct categories: those with integral control gear and those which require separate gear. Since the life of control gear is generally longer than that of a fluorescent lamp, it is often better to install lamps with separate control

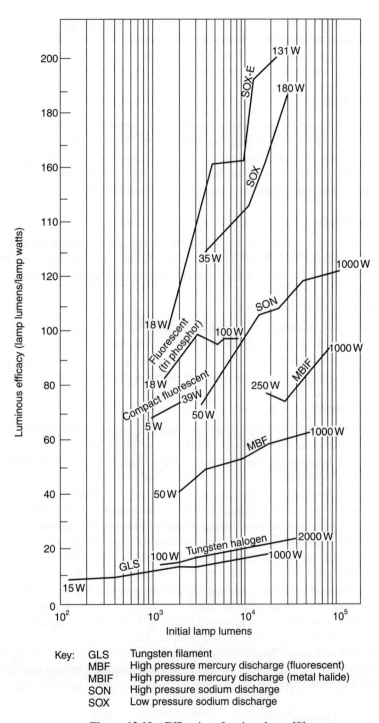

Figure 12.10   Efficacies of various lamps[9]

gear, as this is more cost effective. Lamps with integral control gear are more expensive, being specifically designed as a direct replacement for existing tungsten filament lamps. As with all discharge lamps, compact fluorescent lamps exhibit a poor factor, often as low as 0.5. This however, can be corrected by using capacitors.

Fluorescent tubes

Fluorescent tubes are commonplace in most buildings and exhibit efficacies in the region of 80–100 lm/W [9]. Depending on the type of lamp and ballast used, they can last up to 18 times as long as tungsten filament lamps. In recent years the 26 mm diameter fluorescent tube has replaced the 38 mm diameter tubes as the standard for new installations. These slimmer lamps produce approximately the same light output as the larger diameter lamps, but consume around 8% less electricity [10].

Metal halide lamps

MBI lamps have become popular for a wide variety of applications and are available in a wide range of power ratings, 70–2000 W. They exhibit efficacies in the region of 70–100 lm/W, depending on their power rating [9]. MBI lamps are particularly popular in industrial applications.

High pressure sodium lamps

SON lamps have proved particularly useful for lighting large high bay areas, such as factories and warehouses, where their high efficacy (e.g. 70–120 lm/W) can produce a very energy efficient lighting scheme. They are also useful for exterior lighting, car parks and floodlighting. They are manufactured in a wide range of power ratings from 50 to 1000 W.

12.7.2 Control gear

Because of their nature, all discharge lamps, such as fluorescent lamps, require control gear to operate, which comprises a starting device and ballast. A starting device is required to create the high potential difference between the lamp electrodes, so that an electrical discharge is promoted. Starters can be of a plug-in glow type, which should be replaced every second or third lamp change, or else should be of an electronic type. Ballast is necessary to control the current drawn by the lamp. If ballast were not installed, then the current would increase dramatically as the ionization process takes place with the result that damage would be caused to wiring and the fitting itself.

In addition to the lamp load, ballast consumes electricity. The type of ballast used therefore has an impact on overall energy consumption. Traditionally, ballast has come in the form of a wire-wound choke, comprising a copper wire wrapped around a metal core. Current flowing through the wire produces a magnetic field which dampens the growth of any further current, thus 'choking' the current to the desired level. While conventional wire-wound chokes work very well, they result in excess energy losses, typically in the region of 15–20% of total energy consumption. Recently new high frequency electronic ballasts have been developed. These ballasts run at frequencies between 20 and 40 kHz [10], and can be fitted either with a rapid-start mechanism to facilitate instantaneous starting, or a 'soft-start' which prolongs the lamp-life. High frequency ballasts consume up to 30% less power than wire wound chokes and have the additional benefit of increasing lamp life [10].

12.7.3 Lighting controls

The appropriate use of lighting controls can result in substantial energy savings. These savings arise principally from better utilization of available daylight and from switching off electric lighting when a space is unoccupied. Therefore, when designing a lighting control strategy for any given application, it is important to understand the occupancy pattern in the space, since this will heavily influence the potential for energy savings.

There are four basic methods by which lighting installations can be controlled:

- Time-based control;
- Daylight-linked control;
- Occupancy-linked control; and
- Localized switching.

Time signals may come from local solid-state switches or be derived from building management systems. These signals switch the lights on and off at set times. It is important to include local override so that lighting can be restored if the occupants need it.

Photoelectric cells can be used either simply to switch lighting on and off, or for dimming. They may be mounted either externally or internally. It is however important to incorporate time delays into the control system to avoid repeated rapid switching caused, for example, by fast moving clouds. By using an internally mounted photoelectric dimming control system, it is possible to ensure that the sum of daylight and electric lighting always reaches the design level by sensing the total light in the controlled area and adjusting the output of the electric lighting accordingly. If daylight alone is able to meet the design requirements, then the electric lighting can be turned off. The energy saving potential of dimming control is greater than a simple photoelectric switching system. Dimming control is also more likely to be acceptable to room occupants.

Occupancy-linked control can be achieved using infra-red, acoustic, ultrasonic or microwave sensors, which detect either movement or noise in room spaces. These sensors switch lighting on when occupancy is detected, and off again after a set time period, when no occupancy movement is detected. They are designed to override manual switches and to prevent a situation where lighting is left on in unoccupied spaces. With this type of system it is important to incorporate a built-in time delay, since occupants often remain still or quiet for short periods and do not appreciate being plunged into darkness if not constantly moving around.

Localized switching should be used in applications which contain large spaces. Local switches give individual occupants control over their visual environment and also facilitate energy savings. By using localized switching it is possible to turn off artificial lighting in specific areas, while still operating it in other areas where it is required, a situation which is impossible if the lighting for an entire space is controlled from a single switch.

12.7.4 Maintenance

With the passage of time luminaires and room surfaces get dirty, and lamp output decreases. Lamps also fail and need replacing. Consequently, the performance of all lighting installations decreases with time. It is therefore necessary to carry out regular maintenance in order to ensure that an installation is running efficiently. Simple cleaning of lamps and luminaires can substantially improve lighting performance. Therefore, at the design stage maintenance requirements should always be considered. Luminaires should be easily accessible for cleaning and lamp replacement. Bulk replacement of lamps should be planned, so that they are replaced at the end of their useful life, before light output deteriorates to an unacceptable level. The cleaning of lamps and luminaires should be planned on a similar basis. In order to minimize disruption to staff, planned cleaning and lamp replacement can take place during holiday periods.

References

1. Energy use in offices (1998). Energy Consumption Guide 19, Department of the Environment, Transport and the Regions.
2. Economic use of electricity in industry, Fuel Efficiency Booklet 9 (1994). Department of the Environment.
3. Guidance notes for reducing energy consumption costs of electric motor and drive systems (1995). Good Practice Guide 2. Department of the Environment.
4. Retrofitting AC variable speed drives (1994). Good Practice Guide 14. Department of the Environment.
5. Energy audits and surveys (1991). CIBSE Applications Manual AM5.
6. Smith, B. J., Phillips, G. M. and Sweeney, M. (1987). *Environmental Science* (Chapter 8). Longman Scientific and Technical.
7. CIBSE Code for interior lighting (1994).
8. Pritchard, D. C. (1995). *Lighting* (Chapter 7). Longman Scientific and technical, 5th edition.
9. Energy Management and Good Lighting Practices. Fuel Efficiency Booklet 12 (1993). Department of the Environment.
10. Energy Efficient Lighting in Buildings. Thermie Programme, Directorate-General for Energy. Commission of the European Communities.

Bibliography

Economic use of electricity in industry. Fuel Efficiency Booklet 9 (1994). Department of the Environment.

Energy Efficient Lighting in Buildings. Thermie Programme, Directorate-General for Energy. Commission of the European Communities.

Energy Management and Good Lighting Practices. Fuel Efficiency Booklet 12 (1993). Department of the Environment.

Guidance notes for reducing energy consumption costs of electric motor and drive systems. Good Practice Guide 2 (1995). Department of the Environment.

Pritchard, D. C. (1995). *Lighting*. Longman Scientific and technical, 5th edition.

Retrofitting AC variable speed drives. Good Practice Guide 14 (1994). Department of the Environment.

13

Passive solar and low energy building design

This chapter deals with the use of passive techniques to control the environment within buildings. Through the use of passive solar strategies it is possible to produce an architecture which relies more on the building envelope, and less on the use of mechanical equipment as the primary climate modifier. In particular, the impact of passive environmental control strategies on the design and operation of buildings is discussed. Several active solar technologies are also discussed.

13.1 Introduction

Strictly speaking the term *passive solar* refers to the harnessing of the sun's energy to heat, cool, ventilate and illuminate buildings without the use of mechanical equipment. As with so many artificial classifications it has become somewhat corrupted and now is a generic term for a design philosophy which seeks to produce low energy buildings which are sympathetic to the natural environment. A better term might therefore be *climate sympathetic architecture*, since buildings created by this design philosophy use their envelope as the primary climate modifier and relegate any mechanical plant that is required to a supplementary role. This is in contrast to the twentieth-century practice of erecting buildings with unsympathetic envelopes, thus creating a hostile internal environment, which can only be rectified by the use of extensive mechanical services. While the precise definition of the term *passive solar* architecture may be arguable, there is no doubt that its central aim is to produce low energy buildings which utilize relatively simple technologies. In such buildings the emphasis is on the envelope, with the result that *passive solar* buildings tend to have complex facades, which incorporate features such as external shading, opening windows and light shelves.

While it may be possible in certain applications and in some locations to rely totally on the sun's energy to provide a comfortable internal environment, in most passive solar buildings some mechanical plant is still required. This mechanical plant can be used either:

- to supplement the passive technologies as a secondary climate modifier; or
- as a facilitator, which enables the passive technologies to perform in an optimum manner.

Most *passive solar* buildings are therefore in reality hybrids in which passive technologies are used in tandem with mechanical equipment to achieve a low energy solution. In recognition of this fact, a new term *mixed-mode* has come into being. Mixed-mode buildings are so called because they use a combination of natural ventilation and mechanical ventilation to achieve the desired cooling effect [1]. Mixed-mode strategies tend to provide solutions which are more flexible than those produced by pure passive strategies. They are therefore more suitable for use in speculative buildings where the final use of the building may not be known at the design stage.

Because *passive solar* buildings contain fewer moving parts compared with their mechanical counterparts, it is tempting to believe that *passive* buildings are simpler and easier to design. In fact nothing could be further from the truth. To create a good *passive* building, the designer must have a comprehensive knowledge and understanding of heat transfer and fluid mechanics. Unlike *mechanical* buildings, which use known and easily quantifiable system drivers such as boilers and fans, *passive* buildings use natural 'variables' as drivers, such as solar radiation and wind. This means that considerable analysis must be undertaken at the design stage to ensure that a robust design is produced (i.e. one which operates well under various meteorological conditions). If this is not done, considerable problems can arise and costly mistakes can be made. For example, a building heated purely by solar energy may become uncomfortably cool on days when there is heavy cloud. One major problem for designers is to predict, at the design stage, how passively controlled buildings will behave in practice. Failure to predict performance at the design stage can be a recipe for disaster. Therefore great care must be taken at the design stage. To assist in the design of *passive* buildings, engineers often use complex and powerful tools such as *computational fluid dynamics* (CFD) to predict accurately how such buildings will perform. However, the costs involved in using CFD are high and expertize is scarce. Indeed, in some applications the high cost of CFD analysis and the general lack of expertize in this field are major obstacles to the use of *passive* techniques.

13.2 Passive solar heating

Solar energy is the radiant heat source upon which all life ultimately depends. It is in plentiful supply, even in relatively northerly latitudes. Consider for example, latitudes 45° north and 45° south (i.e. the respective latitudes of Minneapolis in the USA and Dunedin in New Zealand). At these latitudes the noontime solar intensity in mid-winter on south facing vertical glazing is 595 W/m^2 [2]. Given that most of the Earth's population lives between these latitudes, it becomes clear that there is great worldwide potential for harnessing solar energy.

Passive solar heating techniques are particularly well suited to applications which experience both low winter air temperatures and clear skies. Under these conditions the abundant solar radiation available during the daytime can be collected and stored for use at nighttime when heat is most required. Passive solar heating techniques have been successfully applied in North America on many small- and medium-sized buildings [3]. The headquarters building of the Rocky Mountain Institute in Colorado, USA uses solar energy to provide its heating and hot water needs. Through the use of solar energy and a highly insulated envelope, the building manages to maintain an acceptable internal environment with virtually no conventional heating, despite experiencing outside air temperatures as low as −40 °C [4]. Passive solar heating techniques have also been applied successfully in more northerly and cloudier climates. For example, St George's School in Wallasey, constructed in 1961, is an early example of one such building in the UK [4]. However, it is true to say that it is more difficult to utilize solar energy in the temperate and cloudy climates of northern Europe than it is in more southerly climates.

There are four basic approaches to passive solar heating: *direct gain* systems; *indirect gain* systems; *isolated gain* systems; and *thermosiphon* systems [5, 6]. All four techniques aim to store, in various ways, solar energy during the daytime for use at nighttime when outside air temperatures are low. They all involve the use of high mass materials in the building fabric to store heat. In the *direct gain* system, solar radiation enters room spaces directly through large areas of south facing glazing. In the *indirect gain* system, the solar radiation is intercepted by a high mass thermal storage element, which separates the room space from the south facing glazing. The *isolated gain* system is a hybrid of the first two systems, which uses a separate sun space, such as a conservatory or atrium, to capture the solar energy. Finally, *thermosiphons* can be used to promote movement of warm air around buildings.

All solar heating systems rely on the use of large glazed areas to catch the sun's radiation. Glass transmits relatively short-wave solar radiation in the wavelength range 380–2500 nm, but blocks radiation at wavelengths exceeding 2500 nm. Although glass permits solar radiation to enter room spaces, it blocks much of the long-wave radiation which is emitted when the surfaces in the room become hot. This phenomenon is known as the *greenhouse effect* (which should not to be confused with the *greenhouse effect* associated with global warming) and it leads to heat build-up within room spaces. Due to the *greenhouse effect*, the temperature within room spaces rises until the heat losses by conduction and convection equal the heat gains by radiation.

As well as promoting the greenhouse effect within buildings, glazing also plays an important role in the self-regulation of solar heat gains. Glass transmits much more solar radiation when the angle of incidence is small, compared with when it is large. When the angle of incidence is large much of the incident solar radiation is reflected. This quality can be used to great effect by building design-ers. Consider the simple building shown in Figure 13.1. In winter, when solar heat gains are most advantageous, the sun is low in the sky and the angle of incidence on the vertical glazing is small. This maximizes the solar transmission through the glass so that the sun's rays penetrate deep into the room space. Conversely, in summer, when solar heat gains are undesirable, the sun is high in the sky and the angle of incidence on the vertical glazing is large, with the result that much of the solar radiation is reflected. In addition, because the sun's angle of altitude is much higher in summer, ver-tical windows present a much smaller 'apparent' area to solar radiation in summer compared with winter (see Figure 13.2). As a result the solar intensity on vertical glazing is often much lower in summer than in winter. It should be noted that if horizontal roof lights are used the situation is reversed with the greatest solar heat gains being experienced during the summer months.

Most solar heating techniques rely on high mass materials to store heat. A variety of materials can be used to store solar energy; concrete, masonry blocks, water tanks and even rocks have all been used to fulfil this thermal storage role. Essentially, any material which has a high specific heat capacity and is a good conductor of heat can be used in this role.

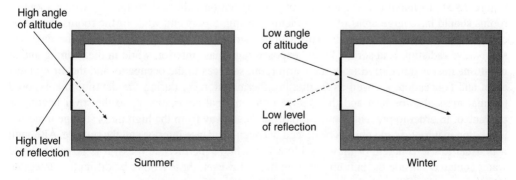

Figure 13.1   Solar reflection from glazing

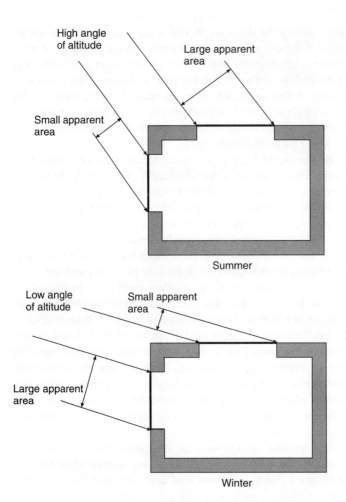

Figure 13.2   Solar angle and apparent area

13.2.1 Direct gain techniques

The utilization of direct solar gains is probably the simplest approach to passive solar heating. It involves using the actual living space within a building as the solar collector (as shown in Figure 13.3). To maximize the amount of solar radiation collected during the winter months, rooms should have large areas of south facing glazing. Floors and walls of the rooms should be constructed from dense materials with a high thermal storage capacity. During the daytime, short-wave radiation is absorbed by the exposed high mass interior, while in the evening and at nighttime heat is transferred from the warm room surfaces to the occupants and the air by radiation and convection. As well as facilitating thermal storage, during the daytime the exposed thermal mass absorbs heat and thus tempers the internal environment, so that over-heating is prevented. In order to prevent conduction of the heat away from the high mass storage material, insulation material should always be placed between the dense interior and the outside. Although it is usual to use concrete or masonry blocks to achieve the thermal mass, it is possible to use water containers inside the building to store heat. However, these tend to be difficult to integrate into the overall building design.

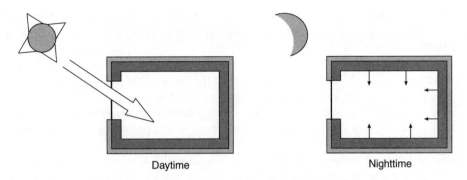

Figure 13.3   Direct gain solar heating

13.2.2 Indirect gain techniques

In an *indirect gain* system, an element with high thermal mass is situated between the sun and the room space. Any solar energy striking the thermal mass is absorbed so that it heats up during the daytime. In the evening and at nighttime heat is transferred to the rooms from the thermal mass by a combination of conduction, convection and radiation. Figure 13.4 illustrates the operation of one such indirect system, the Trombe Wall, which comprises a masonry wall up to 600 mm thick, located directly behind a south-facing glass façade. The outward facing surface of the masonry wall is usually painted black to maximize its absorption of solar radiation. During the daytime solar radiation is absorbed by the masonry wall with the result that the air between

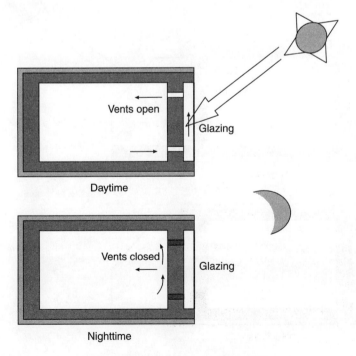

Figure 13.4   Trombe wall operation

the wall and the glass warms up. This causes the air to circulate through vents at the top and bottom of the wall and into the room space, thus warming it. At night the vents in the wall are sealed and the wall transfers heat energy by radiation and convection to the room space.

It is possible to increase the amount of solar radiation collected by a Trombe wall by placing a reflective surface directly on the ground in front of the facade. This material reflects solar radiation onto the thermal storage wall and thus increases its effectiveness.

Trombe walls work best in cold clear climates which experience large amounts of solar radiation, such as those found at altitude in southern Europe. They are much less effective in northern European climates where cloud cover is often extensive in winter.

Another *indirect gain* technique, which has been used in the USA, is the solar roof pond. As well as providing passive heating in winter this system can also provide cooling in summer. It involves constructing a pond on a flat roof. In winter during the daytime the pond absorbs solar energy. At night the warm pond conducts heat through the roof structure and warms the rooms below by radiation. It is necessary at night in winter to cover the pond with movable insulation material. In summer, the pond can be used to provide passive cooling. Overnight the pond is cooled by exposing it to the night air and once cooled, the water mass is used to draw heat from the space below.

13.2.3 Isolated gain techniques

Isolated gain solar heating is essentially a hybrid of the *direct* and *indirect gain* systems and involves the construction of a separate sun room adjacent to a main living space. In the *isolated gain* system, solar radiation entering the sun room is retained in the thermal mass of the

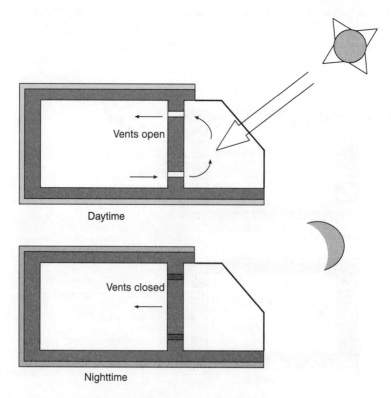

Daytime

Nighttime

Figure 13.5   Operation of an isolated gain system

floor and the partition wall. Heat from the sun room then passes to the living space by conduction through the shared wall at the rear of the sun room and by convection through vents or doors in the shared wall. One classic example of an *isolated gain* system is the use of a south-facing glass conservatory on the side of a house. A typical isolated gain system is shown in Figure 13.5.

13.2.4 Thermosiphon systems

If a flat solar collector containing water or air is placed below a heat exchanger, a thermosiphon will be created. As the fluid heats up in the solar collector it becomes less dense and more buoyant and thus rises to the heat exchanger. As the hot fluid travels through the heat exchanger it cools down and so drops to the solar collector below, where the whole process starts over again. It is possible to heat buildings passively by the use of a solar thermosiphon. In such a system the south-facing solar collectors are placed at a level lower than the room space. Warm air from the collectors is allowed to circulate around a floor void filled with rocks. During the daytime the hot air produced by the solar collectors is used to heat up the rocks and during nighttime the rocks give up their heat by convection to the room space (as shown in Figure 13.6).

13.3 Active solar heating

While passive solar techniques can be used effectively to heat buildings they can place severe constraints on the overall building form and are therefore generally unsuitable for larger more complex building types. However, many of the limitations of passive systems can be overcome by adopting an active system utilizing hot water pumps. Active solar heating systems can be cost effective, provided that they are controlled properly and that pump or fan operating costs are kept to a minimum.

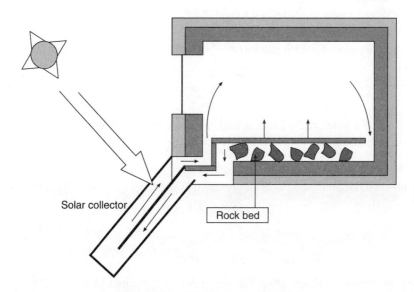

Solar collector

Rock bed

Figure 13.6 A thermosiphon system

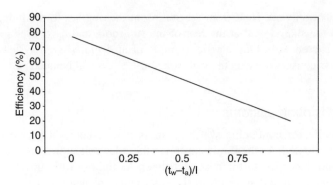

Figure 13.7   Efficiency of solar water collector

The simplest form of active solar collector is probably the flat plate collector, which comprises a 'coiled' metal pipe bonded to a metal plate and placed in a glass-fronted box. In order to maximize the solar absorption the plate and pipe are usually painted matt black; conduction losses are minimized by placing insulation material underneath the plate. The water in such a collector heats up until an equilibrium temperature is reached where the losses by conduction, convection and radiation to the ambient air equal the solar radiation gains. The heat output from solar flat plate collectors can be determined using the Hottel–Whillier equation [7].

$$Q = F[(\tau\alpha)I - U(t_w - t_a)] \tag{13.1}$$

where Q is the rate of delivery of useful energy (W/m^2), I is the intensity of solar radiation (W/m^2), τ is the reflection coefficient, α is the absorption coefficient, U is the overall heat transfer coefficient of plate to air (W/m^{2}K), F is the solar collector efficiency factor, t_w is the mean water temperature (°C), and t_a is the ambient air temperature (°C).

The efficiency of a solar collector is the ratio of the useful heat output over the solar heat input.

$$\text{Efficiency } \eta = \frac{Q}{I} \tag{13.2}$$

$$= F\left[(\tau\alpha) - \frac{U(t_w - t_a)}{I}\right] \tag{13.3}$$

The efficiency of a solar collector is usually represented by plotting efficiency (η) against $((t_w - t_a)/I)$ on a graph (as shown in Figure 13.7).

Example 13.1

Experiments on a flat plate solar collector reveal that

$$F(\tau\alpha) = 0.8188$$

and

$$FU = 7.0041$$

Given this information and assuming that the collector is located in air at 12 °C and receives 400 W/m^2 of solar radiation, determine the efficiency and output of the collector when:

(i) Delivering water at a mean temperature of 40 °C
(ii) Delivering water at a mean temperature of 50 °C

Solution

(i) *At a mean water temperature of 40 °C*

$$\eta = F(\tau\alpha) - \frac{FU(t_w - t_a)}{I}$$

$$\eta = 0.8188 - \frac{7.0041 \times (40 - 12)}{400} = 0.3285$$

$$\eta = 32.85\%$$

Therefore,

$$\text{Output } Q = 400 \times 0.3285 = 131.4 \text{ W/m}^2$$

(ii) *At a mean water temperature of 50 °C*

$$\eta = 0.8188 - \frac{7.0041 \times (50 - 12)}{400} = 0.1534$$

$$\eta = 15.34\%$$

Therefore,

$$\text{Output } Q = 400 \times 0.1534 = 61.4 \text{ W/m}^2$$

Example 13.1 shows that the efficiency of the solar collector reduces dramatically as the mean water temperature increases. This implies that the value of U is not constant and that it is dependent on the water temperature. At higher water temperatures the value of U increases due to increased radiant heat loss from the collector.

It is possible to greatly improve solar collection efficiency by using evacuated tube collectors in which the collector surface is suspended in a vacuum tube. In this type of collector the inside surface of the bottom half of the tube is silvered so that the solar radiation is focused on the collector surface.

Higher water temperatures can be obtained by using solar collectors with silvered semi-circular or parabolic reflectors, which focus the solar radiation of the collector surface. They also reduce radiant heat losses from the collector. With this type of collector it is important that the reflecting surfaces are clean, otherwise efficiency drops off. This can be a particular problem in dry dusty environments, such as deserts.

13.4 Photovoltaics

It is possible to produce electricity from light by utilizing a photoelectric process. The term photoelectric is used to describe any effect which produces electricity from light. There are three main photoelectric processes: photoemissivity, photoconductivity, and the photovoltaic effect. Photoemissivity occurs when materials emit electrons in the presence of light. The photoconductive effect refers to the phenomenon whereby an electric current flowing through a substance is increased as a result of light falling on it. While both these photoelectric processes are used in specialist applications, it is the photovoltaic effect which is most widely used to produce electricity from sunlight. The photovoltaic effect occurs when light falls on the boundary between two substances and causes electrons to be transferred from one side of the boundary to the other. As a result of this transfer of electrons, one material acquires an excess of electrons and becomes negatively charged, while the other loses electrons and becomes positively charged. In this way a positive-negative (P-N) junction is formed. The resulting imbalance in electrons across the P-N junction produces an electromotive force, which when connected to a circuit causes a current to flow.

Photovoltaic cells were first developed in the 1950s in the space industry for use on satellites, but interest in their terrestrial use emerged with the fuel crisis of the early 1970s. Since then, there has been a steady growth in their terrestrial use, mainly limited to remote applications where the provision of mains electricity is prohibitively expensive. In recent years reductions in the cost of manufacture of solar cells has meant that the urban use of photovoltaics has become more popular.

Photovoltaic cells comprise solid-state electronic cells, which are fabricated using crystalline silicon wafers as a substrate onto which another metal is deposited using a screen-printing process [8]. A photovoltaic cell with an area of 100 cm^2 should produce approximately 3.5 A in strong sunlight. Manufacturers encapsulate groups of photovoltaic cells under glass covers to form modules. Within these modules, individual cells are interconnected in series and parallel to produce desired voltages and currents. Similarly, photovoltaic modules can be grouped together to form arrays to increase the power output.

Manufacturers subject their photovoltaic modules to a standard test condition of a solar irradiance of 1000 W/m^2 at an operating temperature of 25 °C, which is approximately equivalent to the solar radiation which would be experienced by a horizontal surface, at noon, in June in Saudi Arabia. It should be noted that the performance of photovoltaic modules drops off as the ambient air temperature increases above 25 °C. This is of particular importance as photovoltaic modules are often used in environments which are much warmer than 25 °C. It has been calculated that operating power reduces by about 0.5% for every 1 °C increase, thus a 100 W module (rated at 25 °C) when operating at 41 °C would actually produce only 92 W.

13.5 Passive solar cooling

The term *passive solar cooling* is a very loose one, which can be used collectively to describe a variety of passive cooling techniques, some of which directly utilize solar energy. However, it is true to say that *passive solar cooling* has more to do with defending buildings against solar energy than utilizing it. Many buildings, especially large commercial buildings, experience over-heating problems during the summer months. These problems often arise because of poor building envelope design. Rather than defending against solar gains, many buildings possess envelopes which actively promote the greenhouse effect, necessitating the installation of large air conditioning systems. There are instead a wide variety of passive techniques which can be employed to prevent over-heating, such as the use of solar shading and stack ventilation. However, in many buildings the use of these techniques alone is not enough to maintain a comfortable environment and so it is necessary to employ supplementary mechanical plant, to provide a *mixed-mode* solution. For example, while a naturally ventilated building may generally experience low levels of heat gain, in specific areas the heat gains may be high and so air conditioning may be required. It is therefore not uncommon to find 'low energy' buildings which exhibit both *passive* and *mechanical* characteristics.

13.5.1 Shading techniques

By far the best way of preventing over-heating during the summer months, or indeed in any part of the year, is to employ adequate solar shading. It is far better to prevent solar radiation from entering a building than trying to deal with it once it has penetrated the building envelope. Shading techniques can broadly be classified as external, internal or mid-pane. External and, to a lesser extent, mid-pane shading techniques offer the best protection since they both prevent solar radiation from penetrating the building envelope. The use of internal shading measures, such as blinds, is much less effective, since although the blinds intercept the incoming solar radiation, they heat up and in time convect and radiate heat to the room space.

External shading can be extremely effective at preventing solar heat gains. By using external shades, such as fins, sails, balconies or even structural members, it is possible to achieve both a 'horizontal' and a 'vertical' shading effect (as shown in Figure 13.8). Vertical shading members, in particular, can be very useful since the sun moves through the sky in an arc from east to west and therefore for most of the time is not directly in front of any one window. However, external shading does have its down side. External shades and fins are exposed to the elements and therefore can deteriorate rapidly if not properly maintained. In addition, in city centre locations they can become colonized by pigeons.

Internal shading usually takes the form of horizontal, vertical or screen blinds which the building occupants can control. Although they are good at reducing instantaneous solar gains, they tend to warm up and emit heat into the room space. A compromise between external and internal shading, is the use of blinds located in the air gap between the two panes of a double window (as shown in Figure 13.9). As the blinds warm up, heat is trapped in the window cavity and relatively little enters the room space. In some advanced windows designs, the warm air produced in this cavity is vented either to outside or to the room space depending on whether additional heating or cooling is required (see Figure 13.9).

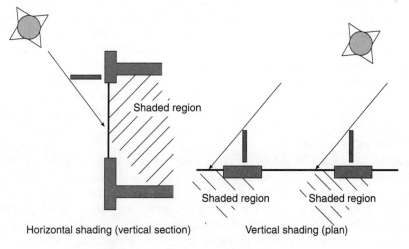

Figure 13.8   External horizontal and vertical shading

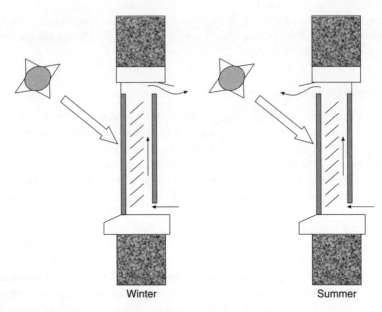

Winter   Summer

Figure 13.9   Mid-pane shading

13.5.2 Solar control glazing

Figure 13.10 gives a breakdown of the component energy flows to and from a 6 mm clear float glass pane. It can be seen that approximately 78% of the incident solar radiation is transmitted directly through the glass, which explains why sunshine has such an instantaneous effect on the occupants of buildings. Approximately 7% of the incident radiation is reflected and the glass absorbs a further 15%. The heat which is absorbed warms up the glass and after a period of time the warm glass emits heat, both inwards and outwards, by radiation and convection.

It is possible to significantly reduce solar heat gains by using solar control glazing. This can be divided into two broad categories, solar absorbing and solar reflecting glass. Solar absorbing glass is body-tinted, typically bronze, grey, blue or green, using a variety of metal oxides. It works in a similar way to conventional sunglasses by reducing the overall transmission of solar radiation through the window. In doing so it also cuts down the transmission of light through the window. Solar absorbing glass exhibits much higher absorption properties than normal clear glass, with up to 70% absorption being achieved with bronze tinted glass [9]. However, it should be noted that although much of the incident solar energy is absorbed, it is eventually re-emitted by the glass, with some of the re-emitted heat entering the interior of the building. Solar reflecting glass is, as the name suggests, highly reflective. The high reflection qualities of the glass are achieved by applying a thin layer of metal oxide to the external surface. Solar reflecting glass can be manufactured in a variety of colours, including silver, bronze, blue, green and grey. The mirrored surface of the glass reflects much of the solar radiation which falls on it and is more effective at cutting down the transmission of solar radiation than solar absorbing glass. It is important to note that this type of glass reflects large quantities of solar radiation and that this can cause problems in surrounding buildings, which may overheat if care is not taken at the design stage.

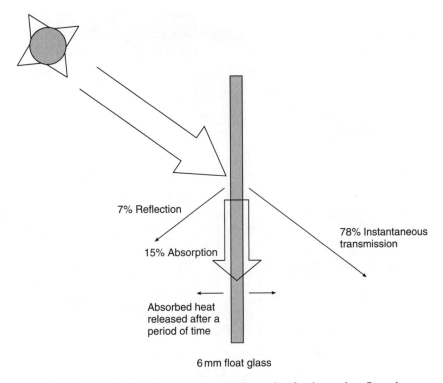

7% Reflection

15% Absorption

78% Instantaneous
transmission

Absorbed heat
released after a
period of time

6 mm float glass

Figure 13.10   Radiation, reflection and absorption for 6 mm clear float glass

13.5.3 Advanced fenestration

One of the characteristics of many passive/mixed-mode buildings is the use of sophisticated
fenestration systems to minimize solar heat gains. Windows in such buildings are often required
to perform a number of different and sometimes conflicting tasks, including:

- Enabling daylight to enter the building
- Promoting natural ventilation
- Promoting solar heating
- Reducing solar heat gains
- Preventing the ingress of noise from outside
- Maintaining building security.

These tasks are usually achieved by installing complex window units, which incorporate some
or all of the following features:

- Solar control glazing (e.g. solar reflecting or absorbing glass)
- External shading
- Internal or mid-pane blinds
- Openable windows or vents.

In addition to the above features, many fenestration systems utilize light shelves to maximize
daylight and minimize energy consumption on artificial lighting. Figure 13.11 shows a typical
advanced fenestration system which might be found in a *passive* or *mixed-mode* building. Such

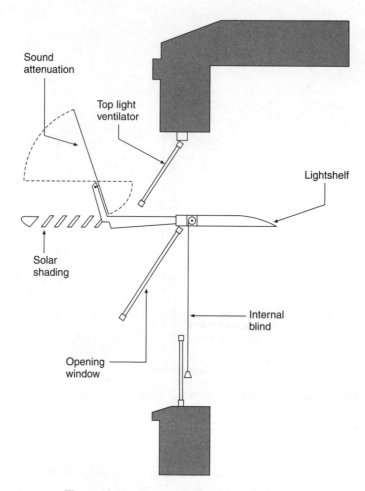

Figure 13.11   Typical advanced fenestration system

fenestration systems are complex and resemble a 'Swiss army penknife'. It is important to appreciate the crucial role played by such windows. In many advanced naturally ventilated buildings the whole environmental control strategy is dependent on the successful operation of these complex windows. If for any reason they cannot be easily operated, the whole ventilation strategy becomes flawed and the internal environment may become uncomfortable.

13.5.4 Natural ventilation

One of the key components of any *passive cooling* strategy is the use of natural ventilation, which can be divided into two basic strategies, cross-ventilation and buoyancy driven (or stack) ventilation. Of the two strategies, stack ventilation is generally more predictable and more reliable than cross-ventilation. This is because, unlike cross-ventilation, stack ventilation is not dependent on wind speed or wind direction, both of which can be extremely variable. The use of stack ventilation is therefore more commonly found in passively controlled buildings than cross-ventilation.

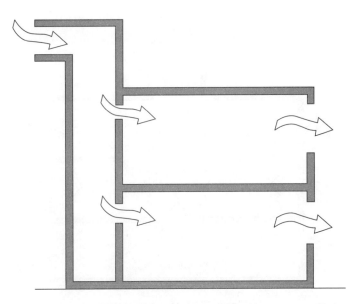

Figure 13.12   The use of a wind-scoop to produce cross-ventilation

Cross-ventilation occurs when openings are placed on opposite sides of a building, so that wind pressure pushes air through the room spaces. As air moves through a building it picks up heat and pollutants, and its temperature rises. This limits the width of room space which can effectively be cross-ventilated. It is recommended that plan width of a cross-ventilated space should not exceed five times the floor to ceiling height [10], which usually results in a maximum width of 14 or15 m. As a result of this, cross-ventilation tends to be restricted to buildings which have narrow plan widths.

Although cross-ventilation is normally achieved by using opening windows, in hot desert countries, wind-scoops are often used (as shown in Figure 13.12). Wind-scoops capture the wind at high level and divert it through the occupied spaces in the building, thus cooling the interior. Wind-scoops can be particularly effective in regions where there is a dominant prevailing wind direction.

Stack ventilation relies on the fact that as air becomes warmer, its density decreases and it becomes more buoyant. As the name suggests, stack ventilation involves the creation of stacks or atria in buildings with vents at high level (as shown in Figure 13.13). As air becomes warmer due to internal and solar heat gains, it becomes more buoyant and thus rises up the stacks where it is exhausted at high level. In doing this a draught is created which draws in fresh air at low level to replace the warm air which has been displaced. Stack ventilation has the beauty of being self-regulating; when building heat gains are at their largest, the ventilation flow rate will be at its largest, due to the large buoyancy forces.

The stack effect is driven by the pressure difference between air entering at low level and air leaving at high level. This can be calculated using eqn (13.4).

$$\text{Pressure difference } \Delta P = gh(\rho_o - \rho_i) \tag{13.4}$$

where ΔP is the pressure difference between inlet and outlet (Pa), h is the height difference between inlet and outlet (m), ρ_i, ρ_o is the density of air at inlet and outlet (kg/m^3), and g is the acceleration due to gravity (i.e. 9.81 m/s^2).

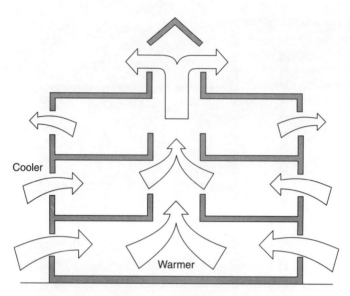

Figure 13.13   Stack ventilation

The density of air at any temperature can be determined using eqn (13.5).

$$\text{Density of air at temperature } t, \rho_t = 1.191 \times \frac{(273 + 20)}{(273 + t)} \tag{13.5}$$

where, 1.191 kg/m^3 is the density of air at 20 °C.

It can be seen from eqns (13.4) and (13.5) that the buoyancy force depends on the:

- Height difference between inlet and outlet vents; and the
- Temperature difference between the internal and external air.

The air volume flow rate drawn by the stack effect can be determined by:

$$V = C_d A_n \sqrt{[(2gh.(\rho_o - \rho_t))/\rho_{av}]} \tag{13.6}$$

where V is the volume flow rate of air (m^3/s), C_d is the coefficient of discharge of openings, A_n is the equivalent area of openings (m^2), and ρ_{av} is the average density of air (kg/m^3).

The equivalent area of the openings (A_n) can be determined using eqn (13.7).

$$\frac{1}{(\Sigma A_n)^2} = \frac{1}{(\Sigma A_{in})^2} + \frac{1}{(\Sigma A_{out})^2} \tag{13.7}$$

where ΣA_{in} is the combined free area of inlet vents (m^2), and ΣA_{out} is the combined free area of outlet vents (m^2).

Example 13.2

A shopping mall is to be cooled using stack ventilation. The mall has vents at low level and in the roof. Given the data below determine:

(i) The pressure difference driving the stack ventilation.
(ii) The ventilation air flow rate.
(iii) The cooling power produced by the stack ventilation.

Data:

Free area of top vents = 12 m^2
Free area of lower vents = 6 m^2
Height difference between vents = 35 m
Mean internal air temperature = 32 °C
External air temperature = 22 °C
Coefficient of discharge of openings is 0.61.

Solution

The equivalent area of openings is determined as follows:

$$\frac{1}{(\Sigma A_n)^2} = \frac{1}{6^2} + \frac{1}{12^2}$$

Therefore,

$$\Sigma A_n = 5.367 \text{ m}^2$$

The density of air at 22 °C:

$$\rho_{22} = 1.191 \times \frac{(273 + 20)}{(273 + 22)} = 1.183 \text{ kg/m}^3$$

The density of air at 32 °C:

$$\rho_{32} = 1.191 \times \frac{(273 + 20)}{(273 + 32)} = 1.144 \text{ kg/mg}^3$$

(i) The pressure difference can be determined by using eqn (13.4).

$$\text{Pressure difference} = 9.81 \times 35 \times (1.183 - 1.144) = 13.4 \text{ Pa}$$

(ii) The volume flow rate can be determined by using eqn (13.6).

$$\text{Volume flow rate} = 0.61 \times 5.367 \times \sqrt{[(2 \times 13.4) \div (0.5 \times (1.183 + 1.144))]}$$
$$= 15.71 \text{ m}^3/\text{s}$$

(iii)   Cooling power $= \dot{m} C_p (t_i - t_o)$

where $\dot{m}$ is the mass flow rate of air (kg/s), C_p is the specific heat capacity of air (i.e. 1.025 kJ/kgK), and t_i, t_o are the internal and external air temperatures (°C).
    Therefore

$$\text{Cooling power} = 15.71 \times [0.5 \times (1.183 + 1.144)] \times 1.025 \times (32 - 22)$$
$$= 187.36 \text{ kW}$$

When employing stack ventilation it is important to remember that the stack effect diminishes the further up the building one goes, because the height difference reduces. The air inlet sizes must therefore increase as one travels up a building in order to maintain the same volume flow rate at each floor level. Because the stack effect diminishes towards the top of a building, it is often worth considering an alternative method of ventilation on upper floors.

Figure 13.14 Queen's Building at De Montfort University, Leicester[11]

13.5.5 Thermal mass

During the 1990s in Europe and the UK a new generation of low energy buildings were constructed, which made extensive use of thermally massive surfaces. Two of the finest examples of these buildings are the Queen's Building at De Montfort University, Leicester [10, 11] (see Figure 13.14) and the Elizabeth Fry Building at the University of East Anglia [12] (see Figure 13.15). These buildings use thermal mass as an integral part of their environmental control strategy to produce a thermally stable internal environment.

Thermal mass can be utilized in buildings to perform three separate, but interrelated, roles:

Figure 13.15 Elizabeth Fry Building at the University of East Anglia[12]

- Mass can be added to the building envelope to create thermal inertia, which damps down the extremes of the external environment.
- Exposed mass can be added internally to create a high admittance environment which is thermally stable.
- Mass can be added either separately or to the building structure to create a thermal store which can be used to cool buildings.

One unique feature of many 'high mass/low energy' buildings is their use of exposed mass to create a high admittance internal environment. In most buildings the widespread use of suspended ceilings and carpets effectively converts otherwise thermally heavyweight structures into thermally lightweight ones, creating a low admittance environment which is poor at absorbing heat energy. Surface temperatures tend to rise in such buildings, making it necessary to get rid of heat gains as they occur. This is one of the main reasons why air conditioning has become a requirement in so many office buildings. However, by exposing the mass of the building structure it is possible to form a high admittance environment, which can successfully be utilized to combat over heating.

The creation of a high admittance environment has implications on the comfort of occupants. It can be seen from the discussion in Sections 8.2 and 11.7 that it is the *dry resultant temperature* and not the air temperature which is critical to human comfort. Provided that room air velocities are less than 0.1 m/s, dry resultant temperature can be expressed as:

$$\text{Dry resultant temperature } t_{\text{res}} = 0.5\, t_{\text{a}} + 0.5\, t_{\text{r}} \tag{13.8}$$

Where t_{r} is the mean radiant temperature (°C), and t_{a} is the air temperature (°C).

By exposing the mass of the building structure it is possible to create a high admittance environment and thus reduce the *mean radiant temperature* and the *dry resultant temperature* of the internal space. Given that the *dry resultant temperature* is the average of the sum of the air temperature and the *mean radiant temperature*, it is possible to allow the internal air temperature to rise in summer without any noticeable discomfort to the occupants, provided that the *mean radiant temperature* is maintained at a lower temperature than would be the case in a conventional building with suspended ceilings.

In terms of thermal storage capacity, floors are by far the single most important element within any building. A 50 mm deep 'skin' of exposed concrete can store in the region of 32 Wh/m^2 °C, giving it considerable potential to provide cooling, if utilized correctly.

Buildings with a high mass envelope are extremely good at reducing peak solar heat gains, because the mass increases the thermal inertia of the building. With heavy masonry walls the time lag between the incident solar radiation occurring on the external face and the heat being conducted to the interior is often in excess of 12 hours. The overall effect is therefore to dampen down the internal diurnal temperature range, thus minimizing peak heat gains. This results in a reduction in the required capacity of any air conditioning plant which may have to be installed.

13.5.6 Night venting

While the use of exposed concrete floor soffits may result in a high admittance environment, problems can still arise if the structure is not periodically purged of heat. This is because the *mean radiant temperature* will steadily rise as the floors absorb more and more heat, until conditions become unacceptable. One effective low cost method by which heat can be purged from a building structure is by night venting. In terms of heat removal capability, ventilation is at its least effective during the daytime, when the difference in temperature between the interior of a building and the external ambient is small. In heavyweight buildings night ventilation is much more beneficial, since the temperature differentials are much greater than during the daytime.

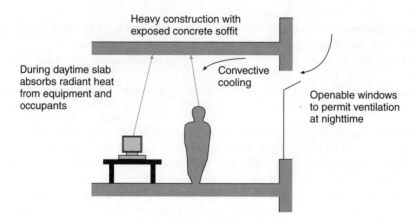

Figure 13.16   Simple night venting scheme in which the windows are opened during the night

Therefore, with night venting it is possible to make the building structure cool, enabling the occupants and equipment to radiate heat to the exposed soffits of the floor slabs.

Night venting involves passing cool outside air over or under the exposed concrete floor slabs so that good heat transfer occurs. This can be done either by natural or by mechanical means. At its most rudimentary night venting may simply entail the opening of windows at night (see Figure 13.16), while a more sophisticated approach may involve a dedicated mechanical night ventilation system and the use of floor voids (see Figure 13.17). If floor voids are used in conjunction with a night venting scheme, then the cool slab can be used to pre-cool the supply air prior to its introduction to the room spaces. In addition, the use of a flood void allows displacement ventilation to be utilized.

13.5.7 Termodeck

When creating a night venting scheme it is important to ensure good thermal coupling between the air and the mass of the concrete floor, and also that fan powers are kept to a minimum. One system which achieves this objective well is the Swedish Termodeck hollow concrete floor slab system.

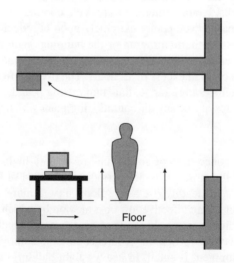

Figure 13.17   Night venting scheme where the ventilation air is introduced through floor void

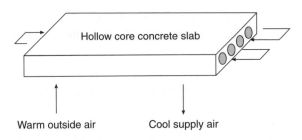

Figure 13.18 The Termodeck system

The Termodeck system has been used successfully in many locations throughout northern Europe and in the UK, notably in the Elizabeth Fry Building [12] and the Kimberlin Library Building at De Montfort University [13]. The Termodeck system ensures good thermal coupling between the air and the building mass by pushing ventilation air through the hollow cores in proprietary concrete floor slabs (as shown in Figure 13.18). By forming perpendicular coupling airways between the hollow cores, it is possible to form a 3 or 5 pass circuit through which supply air may pass, thus ensuring good heat transfer. During periods in which cooling is required, outside air at ambient temperature is blown through the hollow core slabs for almost 24 hours of the day. Overnight the slab is cooled to approximately 18–20 °C, so that during the daytime the incoming fresh air is pre-cooled by the slab before entering the room space. By exposing the soffit of the slab it is also possible to absorb heat radiated from occupants and equipment within the space.

The Termodeck system is particularly worthy of note because it produces buildings which are extremely thermally stable and comfortable without the need for any refrigeration. The example of the Elizabeth Fry Building illustrates this fact very well. In a recent study of low energy buildings in the UK [14], the performance of the Elizabeth Fry Building was outstanding. This building achieved the highest comfort score, while at the same time being one of the lowest consumers of energy; its electrical energy consumption in 1997 was only 61 kWh/m^2 and the normalized gas consumption for that year was 37 kWh/m^2 [12], which compares very favourably with the corresponding values of 128 kWh/m^2 and 97 kWh/m^2 set out in *Energy Consumption Guide 19* for good practice air conditioned office buildings in the UK [15].

13.6 Building form

The decision to utilize a passive environmental control strategy can put severe constraints on overall building form. For example, if natural ventilation is used to promote air movement, it will inevitably lead to the creation of a narrow plan building, unless atria or central ventilation stacks are used. This is because passive buildings are supposed to be climate responsive. Therefore the further an internal space is from an external surface, the less chance there is of harnessing the natural resources of the external environment. The use of a passive solar heating strategy also results in a narrow plan building, but with large areas of south facing glazing. For this reason, passive solar heating schemes tend to be restricted to small- and medium-sized buildings. It is very difficult to use passive solar heating effectively on large deep plan buildings, not least because such buildings tend to overheat for large parts of the year and thus primarily need to be defended against solar heat gains.

Because larger commercial and public buildings generally experience over heating problems, when a passive strategy is applied to the design of these buildings it is usually a cooling/natural ventilation strategy rather than a solar heating one. This means that these so-called *advanced naturally ventilated* buildings all tend to utilize the same generic design strategies and technologies. Broadly speaking these generic technologies/strategies are as follows:

- The use of a heavily insulated outer envelope.
- The use of carefully designed and often complex fenestration, which minimizes solar gains and building heat losses, whilst maximizing daylight penetration. It is also a requirement that the windows can be opened.
- The use of stacks or atria to promote stack ventilation.
- The use of night ventilation to purge the building structure of the heat accumulated during the daytime.
- The careful use of exposed building mass to dampen down swings in internal space temperature, and to promote radiant and convective cooling.

One of the characteristics of these *advanced naturally ventilated* buildings is that they often have complex façades with openable windows, vents, blinds, external shades and even light shelves. These façades incorporate advanced fenestration systems which have many moving parts and which are often controlled by a building management system (BMS). Such complex façades are necessary because the absence of internal mechanical services forces the external skin of the building to become the primary climate modifier and to perform a wide variety of tasks (e.g. day-lighting, defending against solar radiation, ventilation and preventing the ingress of external noise). By creating a complex skin, the designers of such buildings are effectively distributing 'complexity' all around the building rather than concentrating it in a central plant room. This degree of complexity in the skin can have a considerable impact on both the performance of the occupants and the facilities management regime which must be adopted.

In contrast to *advanced naturally ventilated* buildings, the use of a mechanical ventilation system offers considerably more flexibility and enables deeper plans to be utilized. In this respect the Termodeck system appears to offer great potential, as it facilitates good thermal coupling between the air and the building mass without the need for a particularly complex façade or an open plan interior.

The use of the generic passive technologies/strategies described above puts constraints on building design and dictates the building form. With larger buildings the use of natural ventilation often results in buildings which have atria. These buildings comprise a narrow rectangular plan wrapped around an atrium, which gives the appearance of a deep plan building. Vents in the top of the atrium are used to promote buoyancy driven ventilation and air is drawn through vents or windows in the façade. As a result, passively cooled and ventilated buildings tend to look similar to each other, characterized by the use of atria or stacks and complex facades. Figures 13.19–13.22 show sections through four recently constructed passively cooled buildings in the UK: the Learning Resource Centre at Anglia Polytechnic University, Chelmsford; the Ionica Headquarters, Cambridge; the Inland Revenue Headquarters, Nottingham; and the School of Engineering, De Montfort University, Leicester.

It can be seen from the illustrations above that all four buildings exhibit many similarities. In the Anglia Polytechnic and Ionica buildings the designers have used atria to produce buoyancy driven ventilation, whereas in the other two buildings purpose built stacks have been employed. Table 13.1 summarizes the features of all four buildings.

It should be noted that although some of the above buildings, especially the De Montfort building, do not appear to be narrow plan, from a ventilation point of view they are all narrow plan buildings. The practical maximum width of space which can be naturally ventilated is approximately 15 m. This dimension limits the form of the building to a narrow plan format. Nevertheless, by constructing a narrow plan rectangular building around an atrium, which is

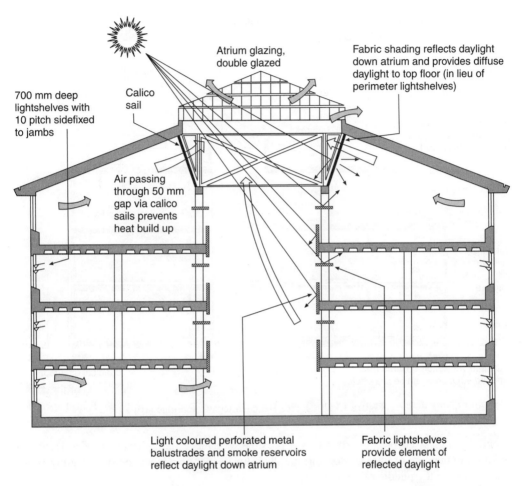

700 mm deep lightshelves with 10 pitch sidefixed to jambs

Calico sail

Atrium glazing, double glazed

Fabric shading reflects daylight down atrium and provides diffuse daylight to top floor (in lieu of perimeter lightshelves)

Air passing through 50 mm gap via calico sails prevents heat build up

Light coloured perforated metal balustrades and smoke reservoirs reflect daylight down atrium

Fabric lightshelves provide element of reflected daylight

Figure 13.19 Learning Resource Centre at Anglia Polytechnic University, Chelmsford [10]

Table 13.1 Summary of characteristics of sample buildings [10]

Building	Atrium or stacks	Complex windows with shading	Night venting	Exposed high mass soffits	Light shelves	Energy consumed per year KWh/m2/year
Anglia Polytechnic University	Atrium	Yes	Yes	Yes	Yes	82
Ionica Headquarters	Atrium	Yes	Yes	Yes	No	64
Inland Revenue Headquarters	Stacks	Yes	Yes	Yes	Yes	89
De Montfort University	Stacks	No	Yes	Yes	Yes	120

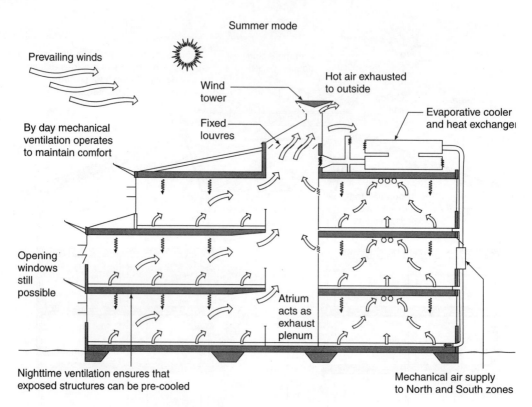

Figure 13.20   Ionica Headquarters, Cambridge [10]

essentially a glass covered courtyard, it is possible to achieve the appearance of a deep plan building. In the case of the De Montfort building, the deeper plan is achieved by putting ventilation stacks in the middle of the building.

13.7 Building operation

The use of a passive environmental control strategy not only affects overall building form, it can also have a considerable impact on the operation of buildings and the performance of their occupants, especially if the building is large and naturally ventilated. By relying on a sophisticated envelope as the primary climate modifier and taking a minimalist approach to the mechanical building services, designers of such buildings need to be careful that they do not create an environment which hinders the performance of the occupants. Although designers may feel that they have produced a comfortable low energy building, the reality may be that it is detracting from the occupier's core business, rather than adding to it. With this in mind, it should be remembered that any property which gets a reputation for being uncomfortable, or unfit for its purpose, is unlikely to gain in value [16].

A recent study of prominent low energy buildings in the UK [14] found that higher levels of occupant satisfaction were easier to achieve in buildings which exhibited:

- A narrow plan form.
- Cellularization of working spaces.
- A high thermal mass.
- Stable and thermally comfortable conditions.

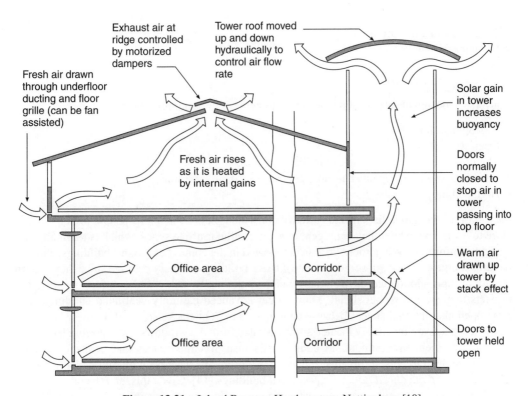

Exhaust air at ridge controlled by motorized dampers

Tower roof moved up and down hydraulically to control air flow rate

Fresh air drawn through underfloor ducting and floor grille (can be fan assisted)

Solar gain in tower increases buoyancy

Fresh air rises as it is heated by internal gains

Doors normally closed to stop air in tower passing into top floor

Warm air drawn up tower by stack effect

Office area

Corridor

Office area

Corridor

Doors to tower held open

Figure 13.21   Inland Revenue Headquarters, Nottingham [10]

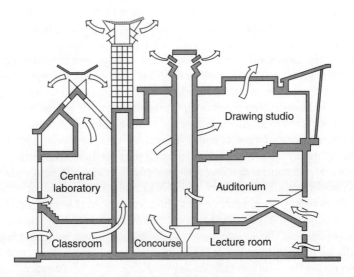

Drawing studio

Central laboratory

Auditorium

Classroom   Concourse   Lecture room

Figure 13.22   School of Engineering, De Montfort University, Leicester [10]

- Control of air infiltration.
- Openable windows close to users.
- A view out.
- Effective and clear controls.

Conversely, occupant satisfaction was harder to achieve in buildings which exhibited:

- A deep plan form.
- Open work areas.
- The presence of complex and unfamiliar technology.
- Situations where occupants had little control over their environment.
- High and intrusive noise levels.

Inspection of the above lists reveals a fairly consistent picture; in short, people prefer to work close to a window, which they can open, in a quiet cellular office space, which is thermally stable and comfortable. When these criteria are viewed in the context of *passive* buildings, a mixed picture emerges. Clearly, in some respects *passive* buildings are very positive since they tend to provide a high admittance environment which is thermally stable and comfortable. In other respects they are not so beneficial. Many larger advanced naturally ventilated buildings have large open plan internal spaces because central atria or vent stacks are used. In these buildings many occupants are inevitably located some distance from the windows and therefore have little control over their environment. In addition, the use of large spaces and acoustically hard surfaces can lead to noise problems. Indeed, noise problems have been highlighted as a particular difficulty in *advanced naturally ventilated* buildings [14]. Given this, and the other reasons mentioned above, it is not surprising that some *advanced naturally ventilated* buildings display low levels of occupant comfort and productivity [14, 17].

The complex nature of the fenestration required in advanced naturally ventilated buildings can be the cause of numerous problems and is therefore worthy of closer inspection. The issue of who has control over the opening of windows is of particular importance. In large naturally ventilated buildings, particularly those which require night ventilation, the function of the whole environmental control strategy can be impaired if, for any reason, certain windows are not opened. Consequently, BMS are often used to control the operation of windows, as the occupants cannot be relied upon to open the windows when required. This automatically, brings the user into conflict with the BMS system, with the result that occupants sitting near windows may be unable to shut the windows when they experience a draught, or conversely, open them when they feel too hot. If a BMS system is not used, issues of conflict can still arise when, for example, the occupants next to the windows may close them in situations where those requiring ventilation in the centre of the building need them to be open. The use of complex fenestration systems, with many moving parts, can also result in poor window sealing over a period of time, resulting in unwanted air infiltration.

From a maintenance point of view, complex fenestration can cause problems. The use of external shading and the numerous protruding and moving parts in these fenestration systems means that they are prone to mechanical damage, and are difficult to clean. The manufacturers of these systems go to considerable effort to reduce potential cleaning difficulties, but it still remains true that from a facilities management point of view these systems need considerably more attention than conventional windows. A study by Kendrick and Martin has shown that the windows most suited for night venting and BMS system control (i.e. high level top hung, and hopper windows) are the most difficult to clean from inside [18]. These advanced fenestration systems have important implications on the flexibility and day-to-day operation of the office space. They are required to be opened by the occupants and are designed to be cleaned from inside. Any desks, bookcases or benches against the external wall inhibit both the ability of the occupants to regulate their environment by opening windows, and the ability to clean and maintain the windows. Consequently, some facilities managers faced with this problem have opted for a 'furniture free' zone next to the windows, or else have installed movable furniture next to the windows. While making the cleaning of windows easy, such a policy can hardly be considered to be designed to increase the comfort and productivity of building occupants. In addition, forcing the occupants away from the windows means that the potential daylighting zone is reduced and thus more people have to rely on artificial lighting, which in energy efficiency terms is very counter productive.

Both developers and building tenants desire buildings which contain flexible space capable of being adapted to meet the evolving needs of organizations. This need for flexibility/adaptability has traditionally been resolved by designing deep open plan buildings. From a facilities management view point the use of an advanced natural ventilation strategy imposes severe constraints on the flexibility of the working space. In particular, it is extremely difficult to partition off internal spaces in such buildings because:

- The insertion of full height partitions may restrict or prevent air movement through the space, thus nullifying its environmental performance.
- It may be difficult to create an acceptable environment within any partitioned office spaces that are created. In a conventional office space it is possible to 'tap' into the nearest mechanical ventilation duct in the ceiling to serve a new space. In a naturally ventilated high mass building there may well be no ceiling, and no mechanical ventilation services in the floor. This makes it difficult to adequately ventilate partitioned spaces.
- The lack of a suspended ceiling can lead to flexibility problems when repositioning luminaires.
- In many advanced high mass buildings the exposed floor soffits are formed by deeply recessed concrete floor beams. The geometry of these floor beams can create problems when erecting partitions.

Even if full height partitions are not installed the environmental performance of a naturally ventilated space may still be impaired by the insertion of high screens and furniture, which restrict the air flow and thus create 'dead' spots.

From the discussion above it is tempting to conclude that all *passive* and *mixed-mode* buildings result in operational difficulties. This however is not the case. The example of the Elizabeth Fry Building, which utilizes the Termodeck system, clearly demonstrates that high mass *mixed-mode* buildings can be very successful. Of all the buildings surveyed in the UK study [14], it was the Elizabeth Fry Building which was outstanding. This building achieved the highest comfort score, while at the same time being one of the lowest consumers of energy. Of particular note is the fact that the building produced an extremely stable thermal environment and comfortable internal temperatures during summer without the need for any refrigeration, clearly demonstrating the success of the fabric thermal storage strategy. The reasons for its success are that the building:

- is thermally stable and comfortable;
- has cellular work spaces;
- has a relatively narrow plan;
- is well sealed and has tight control over air infiltration; and
- has windows which can be opened by the occupants.

These are all qualities which tend to promote user comfort and enhance productivity. When the Elizabeth Fry Building is compared with less successful *advanced naturally ventilated* buildings, it can be seen that its success lies in the fact that the Termodeck system is much more unobtrusive and flexible than the more rigid requirements of the naturally ventilated buildings. For example, the use of mechanical ventilation and hollow core slabs means that the façade of the building can be relatively simple, which frees up the windows so that they can be opened at will by the occupants without impairing the thermal performance of the building. The use of a mechanical ventilation system allows the internal space to be sub-divided into cellular rooms, something which it is difficult to achieve in *advanced naturally ventilated* buildings. It also allows flexibility in the shape and form of the building. Unlike the advanced naturally ventilated buildings where 'complexity' is distributed around the skin of a building, the Termodeck system concentrates 'complexity' in a central plant room where it can easily be controlled and maintained.

References

1. Bordass, W. T., Entwistle, M. J. and Willis, S. T. P. (1994). Naturally ventilated and mixed-mode office buildings – opportunities and pitfalls. CIBSE National Conference, Brighton.
2. CIBSE Guide A2, Weather and solar data, 1982.
3. Hui, H. F., Fong, T. and Lai, K. W. (1996). *Passive solar design in architecture*. The Hong Kong University.
4. Vale, B. and Vale, R. (1991). *Green architecture – design for a sustainable future*. Thames and Hudson.
5. Mazria, E. (1979). *The Passive Solar Energy Book*. Rodale Press.
6. Lebens, R. M. (1980). *Passive Solar Heating Design*. Applied Science Publishers.
7. Francis, W. and Peters, M. C. (1980). *Fuels and fuel technology*. Pergamon Press.
8. Green, M. A. (1998). Photovoltaic solar energy conversion; an update. Australian Academy of Technological Sciences and Engineering, ATSE Focus, No. 102, May/June.
9. Environmental control glasses, Pilkington.
10. Natural ventilation in non-domestic buildings (1997). CIBSE Application Manual AM10.
11. Bunn, R. (1993). Learning curve. *Building Services Journal*. October, pp. 20–5.
12. Standeven, M., Cohen, R., Bordass, B., and Leaman, A. (1998). PROBE 14: Elizabeth Fry Building. *Building Services Journal*. April pp. 37–42.
13. Bunn, R. (1998). Cool desking. *Building Services Journal*. October, pp. 16–20.
14. Bordass, W., Leaman, A., Ruyssevelt, P. (1999). PROBE strategic review: Report 4 – Strategic Conclusions. Department of the Environment, Transport and the Regions, August.
15. Energy use in offices (1997). Energy Consumption Guide 19, Department of the Environment, Transport and the Regions.
16. Beggs, C. B. and Moodley, K. (1997). Facilities management of passively controlled buildings. *Facilities*, **15** (9/10), September/October, pp. 233–40.
17. Leaman, A. (1999). Comfort and joy. *Building Services Journal*. June, pp. 33–4.
18. Kendrick, C. and Martin, A. (1996) Refurbishment: The Natural Way. *CIBSE Journal*, November, p. 29.

Bibliography

Bordass, W., Leaman, A. and Ruyssevelt, P. (1999). PROBE strategic review: Report 4 – Strategic Conclusions. Department of the Environment, Transport and the Regions, August.

Hill, R., O'Keefe, P. and Snape, C. (1995). *The future of energy use* (Chapter 7), Earthscan.

Natural ventilation in non-domestic buildings (1997). CIBSE Application Manual AM10.

Vale, B. and Vale, R. (1991). *Green architecture – Design for a sustainable future*. Thames and Hudson.

Appendix 1

Degree days

A1.1 Heating degree days

The concept of *degree days* was first developed about 100 years ago for use in horticulture [1]. Nowadays, however, degree days are generally used to predict heating energy consumption in buildings. They provide building designers with a useful measure of the variation in outside temperature, which enables energy consumption to be related to prevailing weather conditions.

It is not difficult to appreciate that in a cold month such as January, a given building will consume more heating energy than in a warmer month such as March. This is because:

- the outside air temperature is likely to be colder during January than in March; and
- lower air temperatures are likely to persist for longer in January compared with March.

From this it can be seen that heat energy consumption relates both to the degree of coldness and the duration of that coldness. The degree day method allows for both these factors by setting a base outside air temperature, above which most domestic and commercial buildings do not require any heating. In the UK this base temperature is generally taken to be 15.5 °C. If the average outside air temperature on any given day is below the base temperature, then heating will be required. However, the heat energy consumption in any given period is dependent not only on the magnitude of the temperature differential but also on its duration. For example, if an outside air temperature of 14.5 °C prevails for 24 hours, then a deficit of 1 °C will have been maintained for 1 day and 1 degree day will have been accrued. If the outside temperature remains at 14.5 °C for each day of a week, then a total of 7 degree days will be accumulated. Similarly, if an outside air temperature of 10.5 °C is maintained for 1 week then 35 degree days will be accumulated.

By summating the daily temperature deficits over any given month it is possible to calculate cumulative degree days for that particular month. Therefore, by monitoring daily outside air temperature, it is possible to produce tables of monthly heating degree days for various locations, which can be used by building designers and operators to estimate heating loads. For example, if a particular building experiences 346 heating degree days in January and only 286 in March, it is reasonable to assume that heating fuel consumption in January should be 1.21 times that for March.

Monthly and annual degree day figures are published in many sources. Table A1.1 shows 20-year average heating degree day data for the various geographical regions of the UK.

Table A1.1 UK 20-year average heating degree day data to base 15.5 °C [2]

Region	Jan	Feb	Mar	April	May	June	July	Aug	Sept	Oct	Nov	Dec	Average
Thames valley	346	322	286	205	120	51	22	25	54	130	242	312	2115
South eastern	368	344	312	233	150	74	39	44	82	160	267	334	2407
Southern	345	327	301	229	148	72	39	43	79	150	251	312	2296
South western	293	285	271	207	137	63	28	28	55	116	206	258	1947
Severn valley	321	305	280	201	128	56	24	27	61	138	237	300	2078
Midland	376	359	322	243	162	83	44	48	90	178	275	343	2523
West Pennines	361	340	312	230	144	75	38	39	78	157	267	328	2369
North western	375	345	323	245	167	90	50	56	96	171	284	341	2543
Borders	376	349	330	271	206	117	66	68	104	182	282	339	2690
North eastern	381	358	322	247	168	87	46	49	88	175	281	346	2548
East Pennines	372	352	313	232	154	78	42	44	81	165	272	341	2446
East Anglia	378	349	317	239	149	73	40	39	71	154	269	341	2419
West Scotland	383	352	328	246	170	94	58	64	111	188	299	352	2645
East Scotland	388	357	332	263	197	109	62	67	109	192	301	354	2731
North East Scotland	401	368	346	277	206	120	74	78	127	203	311	362	2873
Wales	330	320	307	240	170	92	49	45	77	145	235	294	2304
Northern Ireland	365	334	320	242	171	92	53	59	99	173	282	329	2519

A1.2 Changing the base temperature

In the UK, degree day data are generally produced for a base temperature of 15.5 °C. However, other countries may use different base temperatures. Indeed, in the UK the National Health Service uses an alternative base temperature of 18.5 °C. It may therefore be necessary to convert data quoted at 15.5 °C to another base temperature. This can be done with relative accuracy by using Hitchin's formula [1] below:

$$\text{Average degree days per day} = \frac{(t_b - t_o)}{1 - e^{-k(t_b - t_o)}}$$

where t_b is the base temperature (°C), t_o is the mean air temperature in the month (°C), and k is the constant.

The value of 'k' varies slightly with location and must be determined from 20-year weather data. However, a general k value of 0.71 can be assumed for most locations in the UK [1].

A1.3 Cooling degree days

Heating degree days are of considerable use when estimating and monitoring the energy consumption of non-air conditioned buildings. However, for air conditioned buildings they are only of limited value. Consequently, the concept of the *cooling degree day* was developed.

Cooling degree days are defined as 'the mean number of degrees by which the outside temperature on a given day exceeds the base temperature, totalled for all the days in the period' [1].

There is, however, no general consensus on the base temperature that should be used for calculating cooling degree days and many users still use a 15.5 °C base [1].

References

1. Degree days. (1993). Fuel Efficiency Booklet 7, Department of the Environment.
2. Energy audits and surveys. (1991) CIBSE Applications Manual AM5.

Appendix 2

Pressure–enthalpy diagram for R22

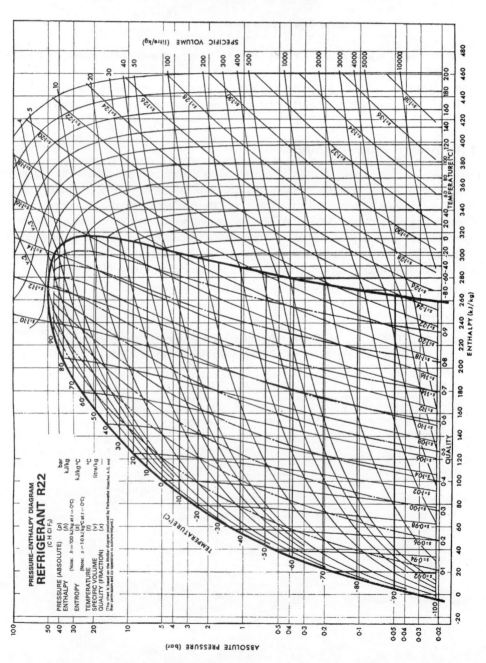

Appendix 3

CIBSE Psychrometric Chart

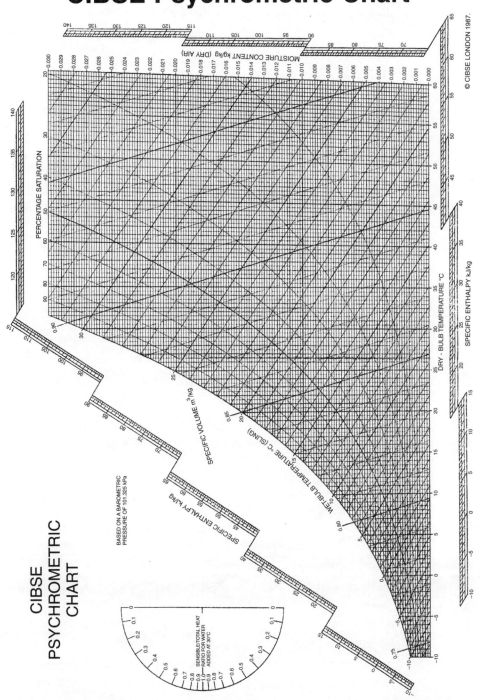

CIBSE PSYCHROMETRIC CHART

BASED ON A BAROMETRIC PRESSURE OF 101.325 kPa

© CIBSE LONDON 1987.

PERCENTAGE SATURATION

MOISTURE CONTENT kg/kg (DRY AIR)

DRY - BULB TEMPERATURE °C

SPECIFIC ENTHALPY kJ/kg

SPECIFIC VOLUME m³/kg

WET-BULB TEMPERATURE °C (SLING)

SPECIFIC ENTHALPY kJ/kg

SENSIBLE/TOTAL HEAT RATIO FOR WATER ADDED AT 30°C

Index